THE HUMAN BAROMETER

A Mother's War, A Planet's Warning

by

Zebedee Njisuh Feka

&

Dr Robinah Namulindwa

Zebedee Njisuh Feka

&

Dr Robinah Namulindwa

Hardcover

ISBN: 978-1-918039-47-4

Good Reach *Publishing*

Dedication

To the mothers — Eliza, Zara, Yasmin, Claire, and Tameka — whose bodies became the barometer.

And to the midwives — Neïma and Mercy — whose hands became the only system that held.

May we finally learn to listen.

Table of Contents

PART I: THE WARNING SIGNS 1

CHAPTER 1 | A GLARING OMISSION 2

CHAPTER 2 | THE FEVER 10

CHAPTER 3 | THE WITHERED FIELD 17

REFLECTION: THE WARNING SIGNS 24

PART II: THE EPICENTER — THE SLOW GLOBAL CRISIS ..27

CHAPTER 4 | THE INCREDIBLE SHRINKING LAKE 29

CHAPTER 5 | WATER AND BLOOD 35

CHAPTER 6 | THE LONG WALK TO KALAMBARI 41

CHAPTER 7 | THE MIDWIFE'S PROMISE 46

CHAPTER 8 | GLOBAL ECHOES — MOTHERS ON THE EDGE OF THE MAP 52

REFLECTION: THE SLOW GLOBAL CRISIS 59

PART III: THE PERFECT STORM, AN ACUTE SHOCK AND SYSTEM COLLAPSE 62

CHAPTER 9 | THE ENDLESS CYCLONE 65

CHAPTER 10 | THE WALLS CAME DOWN 70

CHAPTER 11 | THE ENDLESS CYCLONE 77

CHAPTER 12 | THE ENDLESS CYCLONE 84

REFLECTION: THE PERFECT STORM 91

PART IV: A SHARED PERIL — GLOBAL CONNECTIONS AND CULTURAL SHIFTS 94

CHAPTER 13 | AN UNSPOKEN RISK IN A WESTERN CITY 97

CHAPTER 14 | THE FRAYING ROPE OF TRADITION 105

REFLECTION: A SHARED PERIL 111

PART V: THE FIGHT FOR THE FUTURE 114

CHAPTER 15 | THE ADAPTATION GAP 117

CHAPTER 16 | THE HUMAN BAROMETER 124

CHAPTER 17 | THE HUMAN BAROMETER 131

THE MANIFESTO: THE HUMAN BAROMETER PLEDGE..137

CODA — Listening Again.. 142

Preface

In healthcare, you become a better observer. You have to filter out the noise of a crowded ward to hear the faint, almost imperceptible murmurs in a patient's lungs that signal an impending pneumonia.

You discover that a subtle tremor in a patient's hand, a minor shake in their voice, can be the first hint of a profound neurological storm emerging deep within. The most debilitating illness rarely announces itself with a scream. They begin with these quiet signals, these early warnings. The art of accurate diagnosis is all about attentive listening.

Meanwhile, in the world of climate science, a disparate kind of listening occurs. The planet is the patient, and its whispers are written in the chemistry of fossilised ice cores, in the millimetres of sea- level rise measured by satellites, in the realigning patterns of the jet stream. A doctor spends a lifetime learning to observe the signal within the noise, to identify the tremor in the system that portends a looming, devastating transformation.

We began this journey from these two different worlds, trained in these two distinct forms of listening. One of us, a doctor, was focused on the rhythms of a single human heartbeat. The other, an environmental scientist, was intent on the vital signs of the entire planetary system. Over the years, we worked simultaneously, each of us observing a troubling aggravation of symptoms in our Domain.

It was only when we brought our findings together, placing the fever chart of a pregnant woman in a heatwave next to the temperature graph of the Earth, that we understood the full, terrifying truth. We consider that our analysis didn't involve distinct problems. We were diagnosticians in separate specialities who had just realised our patients shared a single, underlying, and previously unnamed pathology.

Nature was distressed, and its sickness was expressing itself most acutely and dangerously in the bodies of mothers and their developing babies.

In the chapters that follow, we will lay out these first, crucial warning signs as we discovered them. We will take you into the examination room to trace a fever that begins in the atmosphere and ends in the blood. We will walk with you through the withered fields where a changing climate translates directly into a wasting of the human body. And finally, we will expose the most dangerous symptom of all: a glaring, official silence in the very halls of power where this diagnosis should be proclaimed loudly.

"Our patient, the planet, is speaking to us. The first duty of a healer, and of a scientist, is to listen.

| PART I |

THE WARNING SIGNS

A patient is speaking. The most devastating illnesses begin not with a shout, but a whisper — a signal, often missed. As a doctor and an environmental scientist, we have trained our lives to listen. We ask you to join us now, to trace these first, crucial warning signs: a glaring silence in the halls of power, a literal fever in the blood, and a wasting hunger in the fields. This is not a metaphor. It is the intake chart for a planet in distress.

CHAPTER 1 | A GLARING OMISSION

We assumed we had constructed a system where no one would sneak through — a system of protocols, monitors, and policies designed to predict, prevent, and protect. But we had underestimated silence. Not the absence of noise, but the absence of recognition. And in that vacuum, a quiet emergency was already underway, invisible to the very frameworks built to measure risk. It wasn't hidden in shadows. It was under the radar — right where no one had been trained to look.

It's all about sound. It is a steady, reassuring gallop — thump-thump, thump-thump-the sound of a strong heart, digitised, amplified, and broadcast into the quiet, climate-controlled air of

Delivery Suite 7 at St. Thomas' Hospital in London. For Anna, lying on her back with her husband Tom holding her hand, the metronomic rhythm exhibited on the Philips Avalon monitor is a lifeline. It is the digital proof of the life she has so carefully nurtured for thirty-nine weeks, an unwavering reassurance that in this sterile, predictable world, all is progressing according to plan.

All components in the room are evidence of a system scrupulously designed to mitigate risk. An IV line, controlled by an Alaris pump, infuses an isotonic saline solution into her vein to prevent dehydration. The filtered air maintains a constant temperature of 21 degrees Celsius.

A kind, experienced midwife named Chloe makes a small adjustment to the cardiotocograph sensor tied to Anna's swollen abdomen.

"One-forty, perfect heart rate, baby is very happy in there, Anna." She said with a calm counterpoint to the electronic chorus. Beyond, through the triple-panelled sliding window, the fading figures of the Parliament houses are softened by a cool rain of July. In here, there is only the sound of the racing heart, a firm promise of safe passage through a century of medical progress.

Miles away, a round mud-brick hut in a quaint village in the Kanem region of Chad. The wind carried relentless whispers, ochre coloured dust of the Sahel through the open door. Fatima lies not on the hospital bed, but on a woven mat, its Design faded due to overuse. There were no monitors, no machinery, no electronic reassurance. The only technology is the gnarled, knowing hands of the village's matrone, a traditional birth attendant named Yaya. Seventy years old, with eyes that have witnessed more than a thousand births, Yaya occasionally presses a cloth, damp with precious water from a clay jug, to Fatima's forehead.

The only light source available is a single oil lamp that projects dancing shadows on the packed earth walls, making them seem to breathe with Fatima's contractions. Yaya listens, but not to a machine. She listens to the cadence of Fatima's breathing, watches the subtle tension in her jaw. She places a hand, its skin as tough as leather, on the taut skin of Fatima's belly, feeling for the baby's position with a skill born of generations of embodied knowledge. But she couldn't detect the baby's heart rate, and she wouldn't see the subtle declaration that may indicate a compressed umbilical cord. She couldn't measure her blood pressure, driven by dehydration and stress on her body. Fatima's husband, Issa, waiting outside his silhouette, framed by the doorway, his lips moving in silent prayers. Inside, there are only the whispers of the wind and fragile hope for the soul it will soon welcome.

At 2:17 PM, Anna, with a final, expertly trained delivers a healthy baby girl weighing 3.4 kilograms. The room is immediately filled with the infant's lusty, furious cry — a sound of vibrant, well-supported life.

At almost the same moment, 4,500 miles away, Fatima gives birth to a son. He weighs 2.5 kilograms. His first cry is thin, weak, a fragile question mark in the hot, dusty air.

We have seen both of these births. We have stood in the antiseptic calm of the London suite and in the sacred, precarious space of the Chadian hut. The chasm that separates the experiences of Anna and Fatima is ancient, a story of wealth, infrastructure, post-colonial history, and the lottery of geography. It is tempting and convenient to conclude that this is simply a story about poverty. It is not. It is a story about something newer, a global threat accelerator that is actively widening that chasm, dismantling hard-won progress, and making the already dangerous act of giving life in a place like Chad exponentially more perilous. It is a story about climate change. And it is a story whose central, most devastating fact has been almost entirely overlooked by the very people charged with navigating us through the crisis.

After witnessing these scenes and dozens like them, we embarked on a different kind of journey. We left the clinics and the villages and entered the world of policy, of reports, of the grand, global architecture designed to combat the climate crisis. For six months, our small, dedicated team built a database, downloading and analysing the foundational documents of our time: the Paris Agreement, the national climate action plans — known as Nationally Determined Contributions (NDCs) — from over one hundred countries, and the strategic reports of the world's leading development banks. Using keyword searches for terms like 'maternal,' 'newborn,' 'pregnant,' 'neonatal,' and 'reproductive,' we were searching for a single, simple acknowledgement: that the physiological and social chaos of a warming planet poses a unique and severe threat to pregnant women and their newborns.

The stillness we found was ear-splitting. We inspected the NDC of Bangladesh, a country at the forefront of sea-level rise — the plan meticulously detailed initiatives to reinforce coastal barriers and develop rice varieties resilient to seawater. There was not a single sentence on establishing protocols for the safe evacuation of late-term pregnant women from flood zones. We analysed the adaptation fund proposal for a drought-prone nation in the Somali region of Africa. It spoke eloquently of water-efficient irrigation and resilient agriculture, but it was silent on the specific micronutrient deficiencies, such as iron and folate, that prolonged drought inflicts on women — deficiencies that are a direct cause of neural tube defects in their babies and fatal postpartum haemorrhage in themselves. Page after page, the people who are biologically carrying the future of our species were ghosts.

The specific, documented risks — the heat-induced pre-eclampsia, the maternity ward was infected with cholera due to flood, the stillbirths linked to wildfire smoke — were absent. It was not a trivial mistake. It was an intense, systemic silence, a historical relic of institutional design. The modern climate movement, born of environmentalism and energy policy, has always focused on saving forests, on counting carbon, on building seawalls. It simply did not include the word 'stock' or the directive to look inside a delivery room. It was a blatant gap.

This silence is not born of ignorance. The data exists. It is known. Inside the quiet offices of institutions like the World Health Organisation (WHO) and the United Nations Population Fund, a handful of dedicated researchers have been trying to raise the warning for years. We can imagine a man we'll call Dr Aris Thorne, a composite of the real-life, deeply committed officials we spoke to. He sits in a climate-controlled office in Geneva, gazing at a map on his screen, covered in red dots. Each dot is a climate outlier from the past year. He clicks on one in the Sindh province of Pakistan — a heatwave that shattered records. A data window pops up: a 15% spike in stillbirths and neonatal deaths in the month following the heat dome. He clicks another dot over Mozambique — Cyclone Freddy. The linked report contains a desperate plea from a field medic for obstetric supplies and temporary maternity tents. On his desk are printouts of new studies detailing the persistent drought in the Horn of Africa and the increasing frequency of anomalous heat events in the American Southwest, each with its own quiet trail of adverse birth outcomes. He has another pre-publication study from a partner at the London School of Hygiene and Tropical Medicine on his desk. Its conclusion is unequivocal: for every one-degree Celsius increase in ambient temperature during the final week of pregnancy, the risk of stillbirth increases by 6%. He has the proof. The signal is clear.

As the conversation shifted to the upcoming COP conference, Dr Aris Thorne's thoughts turned to the critical data he had compiled. With a sense of urgency, he began presenting his findings, highlighting a pronounced correlation between climate anomalies and adverse birth outcomes. The room fell silent, but the weight of his words was met with a familiar resistance.

The senior director's reply was a masterclass in bureaucratic detachment: "Aris, this is powerful human interest material, truly. But it's a health issue. Our directive here is to emphasise the root alleviation and adaptation levers — energy grids, industrial emissions, carbon pricing." The words hung in the air, a stark reminder of the siloed thinking that had long dominated global policy.

Downstream. The phrases lingered, a perfect distillation of the economic worldview that has controlled global policy for fifty years. It is the cold, clean logic of a system that defines a human life as an 'externality' until it can be assigned a monetary value on a balance sheet. For the man in the expensive suit, Fatima's life-or-death struggle is a "downstream consequence," a rounding error in a global economic calculation. For Aris, it is the entire point. He leaves the meeting with the familiar, bitter taste of institutional inertia. His urgent warning is eventually incorporated, as he feared, into a single paragraph in a lengthy annexe on "Vulnerable Populations," a whisper lost in the institutional roar.

This is how the omission is perpetuated. It is not a conspiracy; it is a failure of imagination, a categorisation error with lethal consequences. It is the inertia of systems designed to see the world through siloed expertise, a direct product of a hyper-specialised global bureaucracy that discourages the interdisciplinary thinking this crisis demands. The carbon experts are not speaking to the experts on childbirth. The result is a worldwide climate strategy that is functionally blind to its most immediate and intimate human consequence.

"*For decades, we have framed climate change as an environmental or economic problem. We have been mistaken. At its core, it is a crisis of public health, and there is no population more sensitive to environmental instability than a mother and child during the 1,000 days from conception to their second birthday.*"

— Dr Maria Neira, Director, Department of Public Health and Environment, World Health Organisation.

The consequences of this omission are not abstract. They are measured in the lives of women like Fatima, whose chances of survival are now shaped not only by the presence of a skilled birth attendant, but also by the number of consecutive failed rainy seasons. They are measured in the futures of babies born with climate-inflicted disadvantages before they even take their first breath. We have chosen, through our collective silence, to prioritise the resilience of our concrete and our coastlines over the strength of the human body at its most vulnerable.

The policy documents in Geneva, London, and New York may be silent, but the bodies of pregnant women are beginning to speak. They are sending an unambiguous signal, a symptom as old as life itself, yet now amplified to a dangerous new pitch by a warming world.

It is the first and most intimate of all the warning signs. It is the fever. But the real cost of omission does not lie in what's missing on paper. It lies in what accumulates in the body — until it burns. Until it breaks. Until the atmosphere outside ignites the crisis within.

CHAPTER 2 | THE FEVER

The body keeps its own weather. Long before a system fails on a grid, it begins to fail in the bloodstream — in the pulse that quickens, in the sweat that no longer cools, in the oxygen that cannot reach where it's needed. A pregnant woman in heat is not just hot — she is entering a state of irreversible banishment, one where her biology becomes a battleground between climate and creation.

The heat does not simply surround the house; it lays siege to it. It presses in on all sides, a physical, suffocating weight. From her place on the sofa, Jessica can hear the low, desperate ambient noise of the air conditioner in the living room, a sound she has come to associate with survival. It is July in Phoenix, Arizona, the third consecutive week of temperatures soaring above 45°C (113°F). Rather than calling it a heatwave, the local news has thrown in the towel, labelling it as new summer standards.

Jessica, seven months pregnant, feels like she is carrying a second sun inside her. Her body has become a furnace, working overtime to cool both itself and the new life nestled beneath her heart. She presses a cold can of seltzer against her temple, a brief moment of relief. Her fingers feel foreign and tight, her ankles are swollen, and a persistent, dull headache has taken up residence behind her eyes. This is just pregnancy, she tells herself for the tenth time that day, repeating the advice from an online forum. Swelling is normal. Feeling tired is normal. But this feels different. A strange, fluttery sensation in her chest, a dizzy spell when she stands up too quickly — it feels like her body is tuned into a machine which is being pushed into a red zone she doesn't have a name for.

Through the gaps in the drawn blinds, the world outside looks colourless and hostile. The asphalt street in her neighbourhood — a part of the city with older, smaller homes and few precious trees — shimmers with a liquid heat, radiating a stored thermal energy that will keep the area dangerously hot long after the sun has set. This is a "heat island," a term she recently learned, a pocket of intense, concentrated heat created by human infrastructure. It means her small rental home is consistently five to seven degrees hotter than the new, tree-lined suburbs on the other side of the city. It is a hidden wall isolating her from a cooler, safer reality.

With a sudden, metallic clang followed by an unsettling silence, the air conditioner gives up — the straining hum stops, replaced by the oppressive, immediate stillness of the heat. The silence is more terrifying than the noise ever was. Within a few minutes, the temperature inside the small house began to rise, the air growing heavy and thick. Panic, sharp and cold, cuts through the lethargy. It is no longer a persistent discomfort; it is an acute crisis. Her body, already at its limit, has just lost its primary defence.

We now know what Jessica was feeling in that moment, not just emotionally, but also physiologically. The "fluttery sensation" in her chest was tachycardia, her heart beating faster and faster to pump blood to the surface of her skin, turning her entire body into a radiator in a desperate attempt to shed heat. This heroic effort, however, comes at a cost. When blood flow to the periphery is reduced, the cardiovascular system is forced to triage, decreasing the flow of oxygen-rich blood to the body's centre — to the kidneys, to the liver, and, most crucially, to the uterus. The placenta, the complex life-support system that she has spent months building, is being slowly depleted of the oxygen it needs to function. The dizziness she feels is a direct signal from her brain that it, too, is feeling the deficit.

Swollen ankles of hers are not just a benign symptom of pregnancy; they are edema, a sign that her blood vessels, stressed by the heat, are leaking fluid into her tissues. The constant headache is a classic sign of severe dehydration, a condition that can dangerously concentrate hormones like oxytocin in the bloodstream, its hormone that initiates labour. In sufficient concentration, it can trick the body into believing it is time to deliver, triggering a cascade of pre-term contractions that can lead to a premature birth, with all its attendant risks of lifelong disability for the child. All of this is happening silently, invisibly, inside a body that is being pushed to the brink. This is the fever, the direct, physiological assault of extreme heat on a pregnant woman. It is an attack that begins in the atmosphere and wages a silent war on a cellular level.

Jessica's predicament is not an anomaly. It is the leading edge of a new and silent epidemic unfolding in developed nations across the globe. While much of the climate conversation focuses on the dramatic visuals of storms and floods, this intimate, internal crisis is escalating rapidly. Phoenix, a modern American metropolis, has become a ground-zero labouratory for this phenomenon. Zebedee, in his work, has studied the mechanics of what makes this city so vulnerable. The urban heat island effect, which traps Jessica in her sweltering home, is a direct result of urban planning choices: the miles of dark, heat-absorbing asphalt and concrete, the lack of tree canopy and green spaces, and the design of buildings that prioritise cost over passive cooling. In essence, we have built cities that function as ovens.

Now, we are placing our most vulnerable populations inside them during unprecedented heat events. These are no longer just "hot days." The heat domes that now settle over regions like the American Southwest are vast, high-pressure systems that are in secure stagnant, scorching air for weeks at a time. They are a direct, predictable consequence of a warming planet and a destabilised jet stream. And as the planet's fever climbs, so does the risk for every pregnant woman living within these new zones of extreme heat.

In a lab at the University of Arizona, a researcher we'll call Dr Lena Hansen is staring at the statistical proof of Jessica's suffering. She is an epidemiologist, part of a small, informal network of global health researchers connected via secure message groups, sharing prepublication data on what they are all beginning to call "climatic-origin obstetric complications." One of her contacts, a UN analyst in Geneva who sounds remarkably like our Dr Thorne, had just shared a briefing on the systemic blind spots in national climate plans, a report that felt chillingly familiar. For the last two years, she has been meticulously cross-referencing emergency room admissions data from every hospital in Maricopa County with daily temperature logs. The pattern is stark and undeniable. On any day where the official temperature exceeds 43°C (110°F), the number of pregnant women admitted to the ER with symptoms of dehydration, pre-term contractions, and hypertensive disorders doubles. If the heat persists for three days or more, it triples. Dr Hansen has identified a clear dose-response relationship: the longer it lasts, the hotter it gets, and the failure of more pregnant bodies starts.

She is trying to secure funding for a simple public health intervention: a targeted alert system, sent via text message directly to pregnant women registered with local clinics, warning them of high-risk days and providing information on the city's public cooling centers. She presents her data to a panel of city health officials. The response is polite, but dismissive. "Dr Hansen, this is fascinating research," says one official, "but we have to be careful about causing a panic.

People in Phoenix know it's hot." Another official, the city's legal counsel, chimes in, his voice betraying the true antagonist. "Besides, our legal department would have a field day with a targeted alert. It implies a level of municipal responsibility we are not prepared to assume."

Lena leaves the meeting with a familiar sense of frustration. They don't see it as a medical event. They see it as a liability. The same glaring omission that exists in global policy documents is happening right here, a failure to connect the dots, driven not just by ignorance but by a politics of liability aversion that prioritizes protecting the city's budget over its most vulnerable residents.

Back in her scorching apartment, Jessica was aware she could not stay. The air is so heavy, she feels she has to chew it to breathe. She thinks of the baby, of the silent, trusting life depending on her for safety. With a surge of adrenaline, she grabs her purse. She looks at her banking app.

There is $237 in her emergency savings account. It's meant for a broken-down car or an unexpected medical bill, not for escaping her own home. But this is a critical situation.

She finds a last-minute deal for a room at a budget lodge off the I-10 freeway, one with a functioning air conditioner. It will cost her $119 for one night. She books it. The expenditure is horrifying, but there was no thinkable choice. It is an act of profound maternal protection, a choice made out of love and fear.

We look at Jessica's choice, and we see both her strength and the system's failure. She found a way to save herself and her child for one night. But her story is a warning. It represents a new, unspoken tax on motherhood in a warming world, paid not just in money, but in stress, in fear, and in the quiet, cellular damage of a body pushed too far. She had the $119. What about the woman in the apartment next door who doesn't?

The fever of the air, the fire in the blood, is the most violent assault. But it is not the only one. This same ruthless heat relentlessly performs a slower, more deceptive sort of assault. It scorches the earth. It thirsted the clouds. It empties the dams. It attacks the body not just with its presence, but with what it takes away: the water that maintains life, and the food that constructs it.

The crisis of the fever gives way to the crisis of the withered field. But if the body burns, it is because something deeper has already withered. And when the rains no longer come, and the soil begins to starve, the harm does not start with the womb — it begins with the land.

Reflection: Clinical Lens

"*We have fundamentally misunderstood what extreme heat is. It is not a weather story. It is a mass casualty event. And pregnancy represents a state of unique vulnerability. The physiological strain of thermoregulation is immense, and when an external heat load is added, the system can quickly become overwhelmed. Every major heatwave is a maternal health crisis in waiting.*"

— Dr Kim Knowlton, Senior Scientist, Natural Resources Defence Council.

CHAPTER 3 | THE WITHERED FIELD

After my own fortress of pride was shattered, I stumbled upon the fragile beauty of brokenness – in the eyes of a mother who'd gone without food so her child could eat, in the worn hands of a father who'd laboured till his body gave out, and in the resilient hearts that beat despite the relentless drumbeat of hunger.

Hunger doesn't always shout itself with a growling stomach. At times, it starts in the blood, in the gradual erosion of eye colour in the steady unravelling of strength that used to feel like ordinary tiredness. Sites, where the rain has stopped writing its story into the soil, the human body becomes the following text of collapse. And in this text, mothers bleed first.

From dawn to dusk, the sound of the grinding stone. It is a soft, pulsating scraping, the sound of Amina's hands pushing a handful of precious sorghum against the worn granite. She is five months pregnant, carrying her second child, and this sound is the metronome of her family's survival. The stack of grain beside her is tragically very small, the last of a ration distributed by an aid truck weeks ago. The flour she produces is hardly enough to cook soor, a thin, watery porridge that will have to serve as breakfast, lunch, and dinner.

Her son, three-year-old Salim, sits listlessly by the doorway of their small, circular hut. He does not ask for the brightly coloured ball made of wound-up rags that he used to love. His eyes, large and luminous in his thin face, follow a lizard skittering across the packed-earth floor, but he does not have the strength to chase it. When the meal is ready, he eats his small portion with a slow, deliberate emptiness, and when it is gone, he looks into the empty bowl and then back at his mother. He does not cry — he has learned there is no more — but his silent gaze is heavier than any scream. It is a look of quiet resignation that no child should ever know. As she watches him, Amina feels a weight settle in her chest, a mix of sadness and exhaustion that seems to grow heavier with each passing day. The fatigue is a physical weight that presses down on her, making every movement feel like a struggle.

The single cooking pot bubbles over a fire made from three carefully chosen acacia wood sticks; fuel, like food and water, cannot be wasted. The water she used to make it, drawn from the community well that morning after a two-hour walk, has a faint, metallic taste. As she stands to rinse the bowls, the world tilts and dissolves into a shimmering darkness. Amina puts a hand against the rough wood of the hut's central post to steady herself. The dizziness passes, but a profound, bone-deep weariness remains. It is a fatigue that sleep does not touch. Later, she pulls down her lower eyelid. The tissue inside is pale, almost white, the colour of bleached bone.

From the doorway, her gaze falls upon the small plot of land beside the hut. It is a graveyard. The maize stalks from the last attempted planting are brittle, ghostly skeletons. Beyond the plot, the cracked earth stretches to a horizon shimmering in the relentless heat. The sky above is a hard, metallic dome, stripped of all cloud. Her husband, Hassan, had sold their last goat the month before. Two weeks ago, their oldest goat had simply lain down under a thorn tree and died. The land could not even support the hardiest of its creatures anymore.

Amina's weariness is more than the fatigue of pregnancy. It is a clinical condition, a profound depletion of the body's life force. Her pale gums and dizzy spells are the classic, visible signs of severe iron-deficiency anaemia, a condition being written at scale across the entire Horn of Africa.

We, as doctors and scientists, have come to understand this condition not just as a personal illness, but as the final, physiological expression of an environmental catastrophe. To understand the danger Amina is in, you must first understand what is happening inside her blood. Imagine your bloodstream as a vast city. The red blood cells are the delivery trucks, and their precious cargo is oxygen. The molecule inside each cell that carries oxygen is haemoglobin. And the essential building block for haemoglobin is iron. Without enough iron, the body cannot build enough haemoglobin. The entire city begins to starve.

A healthy pregnant woman must turn into a nutrition factory, increasing her blood volume by nearly fifty percent to build a placenta and nourish a developing fetus. Her demand for iron doubles. When a woman like Amina becomes pregnant, her body enters a state of profound crisis. Her fatigue is the cry of a body whose organs are being deprived of oxygen. This crisis is an intergenerational sentence. The fetus in her womb is desperately trying to build its own blood supply, but it can only draw from what is available. A child born to a severely anaemic mother is born already in debt, far more likely to have a low birth weight and with minimal iron stores, leaving them vulnerable to infections and developmental delays. The slow violence of the withered field is a shadow cast upon the next generation before it has even taken its first breath.

And for the mother herself, this internal starvation sets the stage for the final, deadly act during childbirth. An anaemic mother has no reserves. Her oxygen-depleted blood has reduced clotting capacity. Her exhausted uterine muscles may fail to contract properly after birth, leading to postpartum haemorrhage. A bleed that a well-nourished woman would survive becomes, for an anaemic woman, a death sentence. This is how the slow violence of a withered field becomes the acute, bloody violence of a maternal death.

The fields are withered because the sky is empty. What is happening in the Horn of Africa is not a simple drought. It is a multi-year, systemic climate failure. The region's lifeblood has always been its two distinct rainy seasons: the Gu rains in the spring, and the Deyr rains in the autumn. But for the past three years, a record five consecutive rainy seasons have failed.

Zebedee has studied the mechanism behind this failure. It is driven by a climate pattern known as the Indian Ocean Dipole. In its extreme "positive" phase, it causes warm, moist ocean air to be pulled east towards Indonesia, while cool, dry air descends over East Africa, catastrophically suppressing cloud formation. Climate models show with chilling certainty that as we heat the planet, these extreme positive

IOD events are becoming more frequent and much more intense. This extended drought triggers a cascade of ecological collapse. The native grasses with their deep, soil-anchoring roots die off. The soil, stripped of its cover, bakes in the sun and becomes hydrophobic — it actively repels water. The ecosystem loses its resilience. The traditional human safety nets for this kind of crisis are breaking. The World Food Programme and other aid agencies, stretched thin by concurrent crises and what they call "Donor fatigue" — a symptom of a world overwhelmed by too many emergencies — are facing unprecedented funding shortfalls. The geopolitics of the moment mean that a slow, grinding crisis in the Horn of Africa is deemed less urgent than a telegenic war elsewhere, and the international aid system, structured around annual budgets, is incapable of responding to a multi-year, attritional disaster. The system is overwhelmed.

Hassan sits outside the hut with his back to the sun, his gaze fixed on the dead land. He closed his eyes, and a sharp and painful memory vividly surfaced. He is a boy of ten running through the same fields after a good gun season, a week before his older sister's wedding. The sorghum in the field is taller than he is. The family has a herd of fifty fat goats. His father has taught him to read the clouds, a skill that now feels like a lost and useless magic. His father is a proud and strong man. Later, placing a hand on his shoulder, his father said, gesturing to the land, The land always provides. A man never leaves his land.

The memory evaporates. He opens his eyes. The green field is gone. His strong father is gone, buried on this land eight years ago, before the rains began to betray them. He rises and walks back into the hut. He looks at his wife's pale skin, at the weary lines etched around her eyes. He looks at his son. And he knows what his father could never have imagined: the land no longer provides. The land now takes.

"*Pastoralist and agro-pastoralist communities in the Horn of Africa have developed sophisticated coping strategies for single-season droughts... But the current multi-year failure is without precedent in living memory. It is not a shock to the system; it is the erasure of the system itself.*"

— Famine Early Warning Systems Network (FEWS NET), Special Report.

The memory evaporates. He opens his eyes. The green field is gone. His strong father is gone, buried on this land eight years ago, before the rains began to betray them. He rises and walks back into the hut. He looks at his wife's pale skin, at the weary lines etched around her eyes. He looks at his son. And he knows what his father could never have imagined: the land no longer provides. The land now takes.

He looks at Amina, and she opens her eyes. An entire conversation — of grief, of fear, of failure, of love — passes between them in that silent, shared look. It is a decision to become a different kind of person. Before they sleep, they perform a final, heartbreaking inventory. They will take the largest water jug, the single, threadbare blanket, and the knife his father gave him. Amina will wrap Salim in her shawl. She will leave behind the small, clay dolls her mother made for her. As she makes this choice, she clutches her belly and thinks of a name she has heard whispered by travellers who have passed through their village, a name of a place where there is help, a place with a midwife known for her skill. The name is Kalambari. The name is Neïma. It is not just a flight from death; it is a walk towards a specific, named hope.

What a person does when the fields turn to dust, and the sky offers no relief, is not complicated: you walk. You walk towards a rumour of life, a whisper of a place where things are better. Across the Sahel and the Horn, millions of people are now being pulled by this desperate gravity.

They are walking towards places like Lake Chad, a body of water that was once a symbol of life, unaware that the same planetary fever that withered their fields has already poisoned the well.

When food becomes dust, and survival becomes movement, people do not walk toward prosperity. They walk toward the last memory of water — even if that memory has already died.

REFLECTION: THE WARNING SIGNS

Here at the precipice of what the Kololo people call Mosi-oa-Tunya, "The Smoke that Thunders," one is struck by the power of a correct description. They did not name this place by its geological cause, but by its observable, sensory effect: the immense plume of mist that rises like smoke and the visceral, earth-shaking roar. They named the symptom. For decades, in our global conversation about the crisis facing our planet, we have been doing the opposite. We have been using a vocabulary of profound inadequacy, a language that obscures rather than reveals.

"Climate change" sounds gentle. "Global warming" sounds cozy. These are anaesthetic terms, words designed to manage panic rather than convey truth. Having examined the initial evidence, we believe these words are now dangerously insufficient. They are a misdiagnosis. The three chapters of this first section are not about an environmental issue. They are the intake chart for a patient in acute distress.

We began with "A Glaring Omission," which is not about policy, but about the psychology of denial. The silence in those global documents is the refusal of the patient's own mind to accept the severity of the illness. It is a convenient denial, one that protects a global economic system that prioritises short-term profit over long-term survival. From there, we moved to "The Fever," the body's most direct response. In the story of a pregnant woman in Phoenix, we see the pathology made manifest. The heat of the planet becomes a fire in the blood. This is not a metaphor. It is pathophysiology. It is the body screaming a truth our policies have refused to whisper. Finally, in "The Withered Field," we charted a more complex, chronic symptom: a wasting syndrome. The withered field and the anaemic woman are the same: a living system's response to a state of prolonged, climactic starvation.

The conclusion is inescapable. At its core, the crisis we face isn't environmental or economic; it's clinical. Climate change is a disease of the planetary system. Maternal mortality is the most sensitive human biomarker. The fever chart in Phoenix, the anaemia in Ethiopia, and the institutional blindness in Geneva are not separate illnesses; they are three distinct readings from a single, failing patient. Viewing the crisis through this lens demands a new kind of response, one of triage, of care, of intervention. It means we must see the pregnant women on the front lines not as statistics, but as the most vulnerable patients in a global ICU.

But to truly understand a disease, a doctor cannot rely on lab results alone. They must go to the patient's bedside. They must travel to the epicentre, to the place where the illness is most concentrated. And on this Earth, there are few places where the symptoms we have outlined converge with more relentless force than the vast, fragile basin of the Sahel. To grasp the full extent of what's happening

| PART II |

THE EPICENTER — THE SLOW GLOBAL CRISIS

A diagnosis is not made from lab results alone. A healer must go to the patient's bedside. For us, that meant travelling to the place where all the symptoms we'd identified converge with relentless force: the Sahel. This is the anatomy of a "slow global crisis" — an illness of attrition the world has found easy to ignore. We will watch an ecosystem wither, see how scarcity births violence, and walk with those forced to flee. The Sahel is not a separate world. It is a bellwether. It is a preview of what is to come.

The river of Zambezi, which thunders over the precipice before us, has not begun its life here. It has a long journey, a slow gathering of strength across thousands of kilometres. It reminds us that every spectacular crisis has a quiet, distant origin. At the falls, the power of the river is an undeniable and deafening spectacle. To understand the river, one must also understand the land of its birth. Every passing year, the lands are growing quieter, hotter and drier.

Travelling to the north, we now leave the thundering spectacle behind for a different kind of landscape in a different kind of crisis. We travel to the Sahel. It is not a country but a vast tapestry of savannah and thorn scrub stretching from the Atlantic to the Red Sea. For millennia, it has been a place of transition, a shoreline of sand against the ocean of green, a landscape defined by its resilience.

But the core we are about to enter is distinct. It does not disclose itself with a single roar. This is a crisis of attrition. A slow violence. If a superstorm is an acute, traumatic event that leaves a visible scar, the crisis in the Sahel is a chronic wasting disease. It is a slow disaster that unfolds over time, eroding the foundations of life so slowly that the world has found it easy to look away. Here, the warning signs we charted in Part I are no longer separate symptoms; they have merged into a single, overwhelming reality.

Our journey through this epicentre will therefore follow the logic of this slow collapse. We will begin at its dying heart, on the shores of a once-great lake. We will witness how its ecological death gives birth to new and terrible forms of human conflict. We will walk alongside those who are forced to flee. And finally, in the dust and heat of a refugee camp, we will find the fierce, stubborn hope of a midwife who refuses to let the future die.

The Sahel is not a separate world. It is a bellwether. The people here are not victims of a unique, isolated tragedy; they are unwilling experts in a future that is rapidly approaching for many more. To listen to their story is to understand the true, grinding, human cost of a planet at war with itself.

Let us enter the epicentre.

CHAPTER 4 | THE INCREDIBLE SHRINKING LAKE

Some absences are chilling, not due to their sudden appearance, but because they were slow. The death of a lake, it's not a single event — it is an uncoiling, a quiet exhale that turns a continent's heartbeat into a hollow echo. You don't grieve at it all at once. You mourn it in pieces: a boat that can no longer sail, a shoreline that has become an illusion, people whose record no longer fits the map.

The lost vintage footage flashes, its colours oversaturated and ethereal. It is 1963, a moment of boundless post-colonial optimism. The camera pans across a vast, glittering expanse of water under a calm and peaceful sky. This is Lake Chad. It is not a lake; it is an inland ocean, the fourth-largest body of water in Africa, a life-giving heart that pumps vitality into four nations.

The screen fills with images of impossible abundance. Great, silvery-white pelicans rise in a flapping chorus. A herd of Kuri cattle — a unique local breed with immense, buoyant horns — wade chest-deep in the shallows. We see the ingenious "polders," fertile plots of land cultivated by Kanembu farmers, a testament to a people who moved with the water's rhythm. The camera finds a fishing boat, a kadey, its hull low in the water, piled high with Nile perch. The board members of the Buduma — the "people of the grasses" — laugh as they pull the nets. On the bustling shoreline at Bol, the air is thick with life: the scents of smoked banda fish, of spices, of wet earth; the sounds of Hausa and Fulani traders, of children splashing, of women dyeing vibrant indigo cloths. It is a portrait of a perfectly functioning ecosystem; it's a symbiosis of thriving; it's a paradise.

Now, a slow dissolve. A dynamic visual of blues and greens bleeds out, replaced by a searing, monochromatic palette of white and ochre. The lush sounds of life fade, swapped by the low, mournful hum of the wind.

On the present day. We are standing on a still floor that was once the floor of that same lake. The camera of our memory has been substituted by our own eyes, and what we see is a panorama of profound absence. The earth under our feet is a fine, grey, salt-crusted dust, cracked into a mosaic of deep, thirsty fissures. There is only a vast, sun-bleached wasteland. The fourth-largest lake in Africa has disappeared, leaving behind the dried corpse of an ecosystem.

To truly perceive a loss of this proportion, one must find a great human anchor. His name is Mustafa, and he is seventy-five years old. His face is a roadmap of deep ageing lines. We find him not with a fishing net, but with a shovel, digging into the dry lakebed to extract clay for making bricks. "Here," he says, stamping a foot. "This is where my uncle taught me to fish. The water was so deep that you could not see the base." He points to a distant cluster of skeletal thorn trees. "And there was the island of Bulama. My wife was born on that island. When we were married, we travelled here by boat. It took half a day." Now, we could walk there in under an hour. Mustafa is a fisherman with no fish, an islander with no water. He is a living archive of a lost world. To stand with a man like Mustafa on the floor of a dead lake, especially while the mist of a place like Victoria Falls still clings to your memory, is to be confronted by a profound question: What killed a body of water the size of a nation?

We have spent years conducting a clinical autopsy of Lake Chad. The cause of death is complex. The most obvious suspect is human thirst. Beginning in the 1970s, the nations of the basin embarked on ambitious irrigation projects, a key part of a post-colonial development model focused on large, centralised mega-projects often funded by international loans. In Nigeria, schemes like the South Chad Irrigation Project (SCIP) were designed to divert enormous quantities of water to cultivate wheat and rice. An official from that era proclaimed with boundless optimism, "We will make the desert bloom." Dams were built on the Chari and Logone rivers, which provided over 90 percent of the lake's water. This was a significant blow, a serious pre-existing condition. But it was not the ultimate cause of death. It was a co-morbidity.

The primary killer was the sky itself. Zebedee's work points to a single, catastrophic event: beginning in the late 1960s, there was a severe and sustained decline in rainfall across the entire Sahelian region. The West African Monsoon began to shift and weaken. This shift has been definitively linked to changes in the surface temperatures of the oceans, a direct consequence of a warming global atmosphere. The planetary fever was starving the Sahel of its rain.

This created a terrifying feedback loop. A large body of water like Lake Chad generates its own local climate, evaporating moisture that falls back in the form of rain. As the lake began to shrink — due to less rainfall and more irrigation — its capacity to generate this moisture. Less water means less rain. The lake, in essence, lost its ability to sweat, and without that evaporative cooling, the entire regional climate grew hotter and drier, further baking the lakebed in a vicious, self-reinforcing cycle of collapse.

The story of this death was not a simple, linear decline. It was punctuated by a period of cruel, false hope. In the early 2000s, a series of years with slightly better rainfall caused parts of the lake to partially recover. This "false dawn" was amplified by the international community. The Lake Chad Basin Commission (LCBC) began promoting an audacious, multi-billion-dollar plan called Transaqua, which proposed to replenish the lake by building a 2,400-kilometre canal to divert water from the Congo River basin. Conferences were held in Abuja and Paris. Headlines proclaimed that Lake Chad was "returning."

We spoke to a family who built their lives on this illusion. Their patriarch, Ibrahim, had moved his family away from the lake in the 1980s. Hearing the news, he convinced his sons to move back. They sold their small home in the city and invested everything in new boats and nets. For three years, their gamble paid off. They built a new home. But the underlying climatic trend had not reversed. The severe droughts turned back after 2010. They noticed as the shoreline they had just settled next to receded again, this time with surprising speed, leaving their boats beached and useless on cracked mud. The loss, this second time, was more bitter than the first. It was the death of a rekindled hope.

"Satellite imagery since the 1S70s provides an unambiguous record of the lake's dramatic contraction... We are not witnessing a cyclical drought. We are witnessing the transformation of a major aquatic ecosystem into a terrestrial one." —NASA Earth Observatory.

We have now conducted the autopsy. The cause of death is clear: a pre-existing condition of unsustainable water management was made fatal by a massive, externally-induced shock of climate-driven drought. The loss, however, goes beyond ecology or economics. To lose a landmark that has defined your civilisation for a thousand years is to suffer a form of collective psychic trauma. It is an ecological grief. The Buduma, the "people of the grasses," are now a people of the dust. Their identity is tied to a world that no longer exists.

But an ecosystem does not die quietly. Its death throes unleash new, savage energies. When the commons vanish, it leaves behind a vacuum. But nature abhors a vacuum, and so does human society. The space left by the retreating water was not filled with peace. It was filled with desperation, with suspicion, and soon, with men carrying guns.

The death of the water was about to demand a sacrifice of blood. But a lake does not disappear quietly. It's vanishing carves new boundaries —

not just of land, but of blood, of rage, of who gets to drink and who gets to live.

CHAPTER 5 | WATER AND BLOOD

Violence does not always arrive with ideology. At times, thirst welcomes it. Long before the first machete is raised or the first home is burned, the war begins at the well — where memory solidifies into doubts, where cooperation frays, and where survival demands a weapon. In a land abandoned by rain and state alike, water becomes the most dangerous border of all.

Before the screaming began, the world was one of quiet routine. Twelve-year-old Omar was helping his father, Bukari, repair a section of their hut's exterior wall. He felt the sense of the wet clay mixed with straw, smooth and cool in his hands. Bakari was teaching him how to intertwine the longer blades of elephant grass into the mixture to create a stronger bond. "A wall is not just mud, my son," Bukari had said, his voice a gentle, low rumble Omar had known his whole life. "It is water, it is grass, it is the work of our hands. All parts must be strong for the house to stand." It was a peaceful, ordinary moment, a gentle memory that is unnoticeable until it becomes the last of its kind.

The sound that broke the peaceful moment was a splintering smash. Not the distant roar of thunder, but an intimate, sharp violence of an axe meeting the wooden door from the neighbour's hut. Omar's hand covered with wet clay, jerked his head up. Outside, the predawn was shattered by quiet. A chaos of aggressive screams exploded, a panicked mix of the Kanuri dialect his family spoke and the distinct, guttural Arabic of the pastoralist community from the neighbouring village. He could hear the panicked bleating of goats and the crushing of mud pots.

Bukari hurriedly grabbed the heavy acacia wood staff he kept near the door. “Stay inside. Do not make a sound,” he commanded. Omar scrambled to a small crack in the mud-brick wall.

What he reached out he see was a scene from a nightmare, the flickering orange light of burning thatch. Men he was long-time associated with were fighting with a savage desperation. They were fighting with their livelihood tools, not with rifles but with farming machetes, heavy wooden clubs, and long handled hoes. He saw his neighbour fall to the ground, his face a mask of disbelief. He saw a boy he played with, perhaps two years older than him, wielding a sharpened stick, his face contorted in a way Omar didn’t recognise.

The chaos swirled around the community’s single, precious well. The well, with its simple stone parapet worn smooth by centuries of ropes, was the only source of water for five kilometres in any direction. Now, it is a battlefield. The air was thick, covered with the coppery smell of blood mixing with the choking dust and acrid smoke. Through the maelstrom, Omar focused on his father. Bakari was defending, using his staff to parry blows. Omar saw a man lunge at his father with a machete. Bukari blocked the blow. Then, from the side, another figure rushed in, swinging a hoe like a medieval axe. The metal blade caught Bukari on the shoulder. He stumbled, his staff falling from his grasp. He went down on one knee.

A scream caught in Omar's throat. His mother pulled him from the wall, her hands shaking. "We must run. Now. Through the back," she hissed. She grabbed a small cloth bundle and pushed him toward the rear of the hut. He scrambled through and emerged into the relative darkness. His mother followed, pulling him towards the dense, thorny bush that marks the edge of the village. He took one last look back. He saw the orange flames consuming the roof of his own home. He saw his father's motionless form on the ground. Then his mother yanks him into the darkness, the only sounds now her ragged sobs and the pounding of his own heart. The smoke that Omar fled from was not the smoke of an ancient tribal war. To mistake it for such is the most dangerous misreading of the new realities of the Sahel. To understand the fire, you must first understand the thirst that lit it.

We have spent time in communities like Omar's, speaking to the elders who remember a different time. The story they tell is not one of perpetual conflict, but of symbiotic coexistence. The two villages at the heart of this tragedy, the Kanuri farmers and the Shuwa Arab pastoralists, had a relationship for generations, based on a sophisticated system of barter and negotiated access to resources. In the dry season, the elders would negotiate grazing rights for the pastoralists' cattle on the harvested Kanuri fields. The cattle would feed on the crop stubble, and their manure would fertilise the land. The farmers traded grain for milk and leather. Their survival was intertwined.

The death of Lake Chad was the cataclysm that shattered this delicate peace. The fishermen among the Kanuri lost their livelihoods. The Shuwa Arab pastoralists lost their grazing lands. Both groups were forced to fall back on the last remaining viable activities: farming small, irrigated plots near the wells and desperately trying to keep their remaining herds alive. Suddenly, they were direct competitors. This is a classic pattern that scientists like Zebedee have seen across the globe, a process known as "resource capture." As a vital resource becomes intolerably scarce, a community may be forced to abandon cooperation and seize exclusive control. This desperation is then often exploited by community leaders, who may engage in "identity framing" — recasting a conflict over water as a conflict between "us" and "them" to rally support.

In this context, climate change becomes a "*threat multiplier*." It takes population growth, existing pressures, and a history of political marginalisation — and intensifies them to an unbearable, explosive degree. The long, slow drought loaded the gun; the dispute over whose turn it was to use the well was merely the pulling of the trigger. This violence was also made possible by the near-total absence of the state. These are "ungoverned spaces," where communities are left to fend for themselves, and where disputes are settled not by law, but by force.

This new, desperate reality created a power vacuum, and it is into this vacuum that a new and far more sinister force entered the Lake Chad Basin: the extremist group Boko Haram and its splinter faction, the Islamic State West Africa Province (ISWAP). Their strategy was as simple as it was diabolical. They arrived as providers, as arbiters, as a new form of authority. A young man named Idris from a similar village. His family was killed in a skirmish over grazing land.

Days later, two ISWAP fighters approached him. They did not preach ideology. They gave him food. They listened to his story. Then, one of them offered him a rifle. "The state will not give you justice," the recruiter said. "But God will, through us. We will give you the power to take back what is yours." For a young man who has nothing to lose, the offer of food, brotherhood, and righteous revenge was an irresistible poison. These groups actively mimic the functions of a state, digging wells, running rudimentary clinics and adjudicating disputes, creating a complex dependency. They turn the climate-driven resource conflict into a recruitment drive for their nihilistic insurgency.

The arrival of these groups transformed the landscape of violence from inter-village fighting into a full-blown insurgency. The state responded with a heavy-handed military presence, often unable to differentiate between civilians and insurgents, deepening the cycle of violence.

Omar now faces this world. Yet, as we have seen in these clash zones, the most significant impacts are reserved for girls and women. In the chaos of a collapsed social order, they face the rampant threats of sexual violence, abduction, and forced marriage to fighters. A woman's body becomes another resource to be captured and controlled. Her survival depends on a form of resilience that is almost unthinkable

In this new kind of war, for every person who picks up a weapon, there are a hundred people — the children, the women, the elderly — for whom there is only one possible choice. It is not to fight. It is to flee. They run from the smoke of their flaming homes, from the memory of blood spilled over a handful of water, from the men with guns. They become a new kind of refugee, a climate refugee, and they begin a journey defined by one desperate hope: to find a place where the fighting has stopped. They begin the long walk.

But when fire consumes the last of what once felt permanent — your family, your fields, your faith in your village — there is no fight left. Only feet. Only flight. And the endless road to somewhere that might still hold life.

Reflection: Political Geography

"It's an undeniable correlation. The rise of extremist violence in the Lake Chad Basin maps almost perfectly onto the areas of the most severe resource scarcity and ecological degradation. These groups are not just ideologues; they are opportunistic political performers who harness climate change. They thrive in the governance vacuums created by environmental collapse."

— Dr Chidi Nwangwu, Senior Analyst, International Crisis Group.

CHAPTER 6 | THE LONG WALK TO KALAMBARI

Someone who stayed, stayed too long, believed too much, loved too hard and began to become a refugee. And as their belief is buried to ash, what's left is not just displacement — it's a moving body, carrying a future that grows heavier with each pace. To walk pregnant through a disappearing world is to perform the impossible: gambling of life to the rumour of safety.

There is no time for grief. Grief is a luxury, a quiet space that no longer exists. It is a room in a house that is now a pyre of ash. There is only the tearing sound of her own breath and the sting of smoke in her lungs, a scent of burning thatch and something else, something terrifyingly sweet — the smell of the village's grain store, their entire future, turning to burned carbon. She's Zara. Six months ago, she was a wife and a daughter, a weaver of intricate patterns into her clothes. Now, she is only a pair of running feet.

The thorns of the acacia scraps tear at her ankles, leaving fiery, thin lines on her skin, but she can't sense them. Her world has diminished to the small patch of ground immediately in front of her, illuminated by the hellish orange glow of her village burning behind her. All faded screams are replaced by the blaze of fire and the pounding in her own ears. She clutches the small, hard lump of her belly, a reflexive, protective gesture. It is the only solid thing in a world that has turned to smoke and terror. Forward is the only thought her mind can form. Just forward.

She runs until her legs are hard and lungs are fire. She stepped into a sandy, dry riverbed, the wadi that had been lifeless for two years, and scrambled up the other side, finally colliding into the blackness of a rocky outcrop. Her energy recedes, and in its place, a crushing, physical exhaustion floods her body. Shakingly, she tightly curls herself into a ball. And now, in the sudden quietness of the Sahelian night, grief comes for her. It is not a soft sorrow. It is a violent, silent convulsion. A memory flashes behind her eyes: her husband, Musa, laughing as he lifted a full fishing net from the water. Another follows: her mother, pressing a heavy, small cloth bundle into her hands just before forcing her through the back of the hut, her last word a fierce, desperate command: "Survive." Zara now opens the bundle for the first time. Inside are some silver coins from her dowry, and a smooth, small cowrie shell her mother had always kept for luck. She can still feel the warmth mother's hand. She tightly grasps it, her body wracked with a sadness so profound it has no sound. There are no tears. Her body does not have the water to spare

The sun rises with violent speed. The panicked flight is over. The long walk has begun. Zara gets to her feet, her joints stiff, her body a single, unified ache. She has no destination. Her only direction is away. Her skeletal shadow begins to walk towards the south. This journey quickly settles into a monotonous routine. She wakes early and starts walking in the relative cool, rests during the midday, and walks again as the afternoon descends, until darkness makes it horrifyingly dangerous. She inspects her feet, now full of blisters and raw skin, and re-wraps them with pieces of cloth torn from her headscarf.

The landscape she moves through is an alien world, a place scoured of life. The domain of Zebedee's science, made terrifyingly real. The ghost forests of dead acacia trees, their bark peeled away, stand like silent, accusing sentinels. The earth is a hard as baked clay, so small that even if rain were to fall, it would simply sheet off. One afternoon, she stumbles upon another village. It has not been burned.

It is simply empty. The huts are intact, the doorways gaping open. A child's sandal lies half-buried in the dust. It is a ghost village, a silent testament to a community that saw the end coming and simply walked away. The quiet here is more chilling than the screams of her own village. It is the silence of surrender.

Her body, the world that Dr Robinah understands with such intimacy, is beginning to send its own desperate signals. Thirst is constantly present. Hunger is a duller, but persistent ache. She took a small handful of roasted groundnuts, three in the morning and three at night. And then there is the baby. It has its own rhythms. It is quiet when she walks, lulled by the motion, but when she rests, it begins to move. A kick, a roll. These movements are her only companions. "We are resting now, little one," she'll murmur, a hand on her belly. "You must be strong." Every kick is both a reassurance — I am alive, I am still here — and a reminder of her terrifying responsibility.

One night, a sharp, cramping pain occurs in her lower belly. Terror grasps her. Is it a miscarriage? Is it labour? Dr Robinah would recognise them as Braxton-Hicks contractions, made intense by dehydration and stress. Zara only knows them as a terrifying portent. The cramps eventually subside, but leave her with a new, heightened sense of her own fragility. She is a walking, breathing, high-risk pregnancy with no help, no clinics, no midwife.

Weeks blur into a single hope of rhythm. One day, she sees another family. The older woman sees Zara's swollen belly and, without a word, gives a single, dried date, pressing it into Zara's hand. The intense sweetness is a shock, an act of generosity so profound it almost breaks her. It is from this family called "Kalambari."

A camp, they say. Near the big city, N'Djamena. There is water, food, and doctors. For Zara, the name becomes a focal point. It is no longer a directionless flight. It is a pilgrimage. Kalambari. The word is a shield against the pain in her feet and the fear in her heart. She turns her face southeast and begins to walk with a new, desperate purpose.

The landscape begins to show signs of a great migration: a discarded cooking pot, a broken wheelbarrow. It is a graveyard of abandoned lives. The hardest push was the last one. After what feels like a lifetime, she senses a change. She smells the distant, faint scent of thousands of woodsmoke fires. Then she hears it: a low, continuous sound of a large crowd of people.

As she reaches the top of the barren hill, the scene unfolds, and there it is. It's not an oasis, but an extended, chaotic city of white and blue tents, shining in the sun. The scale of it is overwhelming. She steps forward, her legs giving up and can no longer hold her weight, and she collapses at the camp's corner. The sensory overload hits her: the smell of cooking fires and waste, the sound of different languages. She is just one more arriving in a ghost city. She's all covered in dust, her lips cracked and bleeding, her clothes in tattersAs she fell on the ground, a woman appeared before her, same aged as hers, wearing a red vest. Her own tiredness wrinkled on her face, but her eyes are kind and clear. The woman kneels in front of Zara, placing a steadying hand on her arm, and speaks in a gentle voice. It is the first kind touch she felt from a stranger since she left her home. In that moment, the only thing standing between life and death were the unfamiliar women words. She only knows she has arrived.

But even arrival does not equal rescue. What meets you at the end of the road is not always relief, but a different kind of test — one where the tools of survival are gentleness, trust, and hands steady enough to stop death mid-breath.

Reflection: Displacement as Clinical Collapse

"*Displacement is a public health emergency. For a pregnant woman, it represents a catastrophic failure of all support systems... Every kilometre she walks, the statistical probability of a safe outcome for her and her child decreases.*"

— Statement on Forced Displacement, United Nations Population Fund (UNFPA).

CHAPTER 7 | THE MIDWIFE'S PROMISE

In the fading light of the tented city of Kalambari, where the sounds of war and displacement fade into the background, and the cries of new life rise above the din. For Neïma Patricia, a midwife with a steadfast heart, every birth is a defiant celebration of hope in the face of uncertainty.

In the unfolding world, not all heroes wear stethoscopes. Some hold memory, respect, and the quiet courage to kneel beside an elder and say, "Guide me." The only bridge strong enough to hold a mother through labour and a baby through breath is the space between a grandmother's suspicion and a midwife's open hands. This is where policy ends, and promise begins.

The day is ending at Kalambari camp, but the work never does. In the main UNFPA-supported maternity tent, a structure of white canvas and blue logos that has become her entire world, Neïma Patricia is finishing her notes. The air inside is still thick with the day's oppressive heat, but it is a space of meticulous order in a world of chaos. Clean instruments gleam on a sterile blue cloth atop an overturned supply crate. A general ledger, filled with her neat and precise handwriting, records every birth, every complication, every life brought safely into this sprawling city of tents. This ledger is her testament. As she writes, her mind drifts for a moment to the sterile, air-conditioned hallways of the hospital in

N'Djamena where she trained. She could have stayed there, built a comfortable career. But she also remembers a cousin of hers, Amara, from her home village, who died during childbirth, bleeding to death on a dusty floor for want of a simple dose of oxytocin that cost less than a cup of tea. That memory is why she is here. Her work in Kalambari is not a job; it is a promise she made to a ghost.

Her last consultation of the daylight hours is with Zara, the woman who walked from the ashes of her village. Neïma has cleaned her feet and given her a high-energy nutritional paste and oral rehydration salts. Now, she performs a gentle examination. As her practiced hands move over Zara's six-month-pregnant belly, she is conducting a rapid, silent triage, compiling a mental risk score.

Gestational age approx. 28 weeks + visible malnutrition (low BMI) + clinical signs of dehydration (poor skin turgor) + probable psychological trauma = high risk for pre-term labour and placental insufficiency. "You are safe now," Neïma says, her voice calm and even. "You and your baby. We will take care of you both."

Just as she is helping Zara to a clean mat, another woman rushes in, her face a mask of panic. "Neïma! It is my sister, Femi. The baby is coming now, but my grandmother… Maman Binta… she will not let her come to you. She says the old ways are best."

Neïma feels a familiar tightening in her chest. She takes a deep breath, her gaze steady, and picks up her small leather bag. "Show me where she is," Neïma says, her voice leaving no room for argument. "We will go to her."

Understanding Maman Binta's psychology, which may appear stubborn, but for her, it's about holding on to the vital.You have your traditions. They are the last solid ground in a world that has turned to quicksand. The "official" world — the government soldiers who failed to protect them, the aid agencies with their endless paperwork — has not always proven trustworthy. Maman Binta is not a villain; she is a woman holding on to the last thing she believes she can control.

Neïma walks through the winding, chaotic lanes of the camp as dusk settles. She arrives at a small cluster of shelters where the old woman, tall and resolute, stands guard. Neïma does not approach with aggression. She respectfully keeps a distance and greets the matriarch in her language. "Maman Binta, I have come because your granddaughter asked for me. I am here to lend a hand."

The old woman's eyes are like chips of obsidian. "We do not need your help. The women of my line have always given birth with the help of our ancestors."

This is the critical moment. "That is true," Neïma says, her tone one of genuine respect. "And your ways are strong. My own grandmother was an attendant in our village. She used the leaves of the neem tree for fevers." This shared heritage hangs in the air, a small offering. "She was a wise woman. She also taught me that sometimes, the old ways need a new helping hand. I have heard you are wise in these matters as well. Tell me what you have done for Femi already."

Maman Binta, taken aback by the deference, hesitates, then speaks of an herbal tea. Neïma listens patiently. "Your wisdom is great, Maman,"

Neïma says when she is finished. “We are both on the same pace, we both want a healthy mother and a healthy baby. Sometimes, the baby is like a traveller who has lost its way on the path. My training has taught me some extra ways to see if the path is clear.” She opens her bag slowly. “May I hear the heartbeat of the baby with my trumpet? It will not disturb you. We can work together, your wisdom and my training.”

The offer of collabouration hangs in the air. After a long, tense silence, the old woman gives a single, sharp nod and steps aside. Neïma enters the dark, hot tent. The labouring woman, Femi, is on her knees, her moans deep and guttural. Neïma’s calm demeanour vanishes, replaced by a laser-like clinical focus. Her diagnosis is swift and certain: The baby is in a transverse lie — sideways. It is an obstructed labour. Without intervention, both mother and baby will die.

Working with a speed and efficiency that seems almost supernatural, Neïma becomes a brilliant physical strategist, using her hands, her knowledge of anatomy, and the mother’s own body to manually guide the trapped fetus. Finally, with a great rush, a healthy baby boy is born. But Neïma’s work is not done. She stays for another hour, checking for haemorrhage, helping the new mother latch the baby for its first feeding. Maman Binta watches from the entrance, her silence no longer hostile, but filled with a dawning awe. It is this quiet, patient after-care that seems to affect her the most. As Neïma is finally packing her bag, the old woman speaks. “The tea I gave her… it was to make the birth faster. Was that wrong?” It is the first crack in the armour of certainty “You did what you knew to help," Neïma replies gently. "But sometimes a fast birth is not a safe birth. Tomorrow, if you like, I can show you the signs to look for."

This is Neïma's promise, made manifest. It is not just a promise to deliver babies. It is a promise to show up, to listen, to respect, and to build bridges of trust where they have been burned away. It is because of this promise that here, in this small section of the vast Kalambari camp, under her team's direct care, not a single mother has died in childbirth. Her skill and diplomacy create an island of safety against the slow, grinding crisis of displacement.

But what happens when the crisis is not slow? What happens when it does not arrive on foot over many weeks, but in a matter of hours, with the force of a falling sky? The resilience built in a place like this is a testament to the human spirit. But a different kind of storm is gathering, one that threatens to shatter even these hard-won sanctuaries.

The war against a slow crisis is about to collide with the fury of a perfect storm.

But no promise, no matter how sacred, can hold back every storm. Because some storms don't arrive on foot or through silence — they arrive with a roar, crossing oceans, borders, and air. And they find mothers even at the edge of the map.

Reflection: Trust as a Clinical Tool

"*Effective maternal healthcare in displaced communities is not about replacing traditional practices but about integrating with them. The midwife's role is as much cultural as it is clinical.*

Trust is the most vital clinical tool we have."

— Dr Olajumoke Adebayo, Sociologist and author of Birth in the Age of Displacement.

CHAPTER 8 | GLOBAL ECHOES — MOTHERS ON THE EDGE OF THE MAP

You can change the language. You can change the landscape. But if you follow the symptoms — salt in the drinking water, smoke in the lungs, silence in the ultrasound room — you'll find the same condition. It is not a regional issue. It is a global pathology, written in the breath of women whose bodies are bearing both life and collapse at once.

The crisis we witnessed in the Sahel was not a local tragedy. It was a diagnosis. And diagnoses, once spoken, echo. They travel through bloodlines and coastlines, crossing oceans, morphing into new symptoms. But the condition remains the same: systems that fail to protect mothers in the face of a heating, hardening world. This chapter follows four such echoes — across rivers, heatwaves, smoke, and silence — to remind us that no geography is immune, and no womb exists outside the storm.

Yasmin crouched at the edge of the pond, with trembling fingers, she filled a rusted tin with water that shimmered strangely in the morning light. She had drawn water here all her life, but now it tasted like seawater. Metallic. Faintly bitter. The embankment had been torn open by the cyclone last year, and the river, swollen with salt, had never fully receded. The water they drank, the water she bathed in, the water she used to cook lentils for her son — none of it was truly fresh anymore.

With her second child, she was eight months pregnant, her body already swollen from the heat and weight of carrying. Her ankles had vanished beneath tight skin. Sometimes she couldn't catch her breath and felt dizzy every morning. The nurse at the Shyamnagar health post mentioned the word pre-eclampsia, but there was nowhere nearby that could test her blood. Alim, her husband, had no money for transport to the district clinic, and Yasmin didn't want to cause alarm. So, she stayed home, lying under the ceiling fan that barely stirred the air and drinking what they had.

She blacked out one morning as she was getting up from her haunches. Her hands were moist, and her mind was spinning when she woke up on her hut's floor. She was seizing when they reached the clinic. Hours later, her daughter was born by emergency caesarean.

She was tiny, breathing with effort, and wrapped in someone else's scarf. They told Yasmin she had a condition brought on by high sodium levels. That she needed rest and proper hydration. But there was no bottled water in the village. Only the pond. Only the salt. And a map that pointed adaptation funding toward sea walls and shrimp farms, not the blood pressure of women like her.

Hands darkened with ash, María stood at the edge of the clearing. Licking at the edge of the forest, the fire had come closer in the night, where the illegal miners worked. She'd grown up fearing snakes and jaguars, not smoke. But in the last year, the dry season in Madre de Dios had stretched too long, and the sky rarely rained anymore. Her second child grew inside her, restless, shifting low in her belly as if searching for cooler ground.

Each day, she breathed in more haze. She used a wet cloth over her face, but it did nothing. A wheeze developed that wouldn't go away, and her chest tightened. There were no masks, no oxygen in the health outpost near the river, no one who could explain what wildfire smoke did to a body, making another body. She felt guilty watching her toddler cough in the mornings.

They stopped catching fish from the stream; the mercury from the gold washings had made even the smallest ones turn bitter.

At 34 weeks, she woke coughing up blood. Her husband paddled her downriver for hours, the heat rising from the scorched banks like a fever. The baby came early, small, and struggling to breathe. They told her he would need to be monitored for lung complications. She nodded without understanding. The baby had survived, but she had the quiet sense that something inside her had not.

No one mentioned the fire. No one blamed the smoke. No one traced the particles in her blood back to the policies in Lima that turned a blind eye to illegal extraction. María returned home to a forest still burning. Her son slept lightly, waking to every crackle in the night.

Camille pressed her back against the tiled wall, the cool of it barely cutting the Marseille heat. It was day three without relief. The air inside the apartment didn't move. The fan whirred, then stopped. Her appetite lost to nausea and fatigue; she hadn't eaten anything since yesterday. Her partner, Karim, laid soaked towels across her belly, her chest, her feet. The heatwave was breaking records, but in the quarter where they lived, no one came to check. No one brought supplies. There were no trees to shade the concrete. No sea breeze reaches their floor.

To avoid overheating, her midwife had warned her to stay hydrated. But the tap ran warm, and the lights flickered. At thirty-six weeks, Camille stopped feeling the baby kick. At first, she told herself it was the heat making her slow, making her imagine things. But when she pressed her belly, and nothing moved, she knew.

The hospital was full. The staff moved in slow motion. The scan showed stillness. The baby's heart had stopped sometime during the night. "Unexplained," they wrote.

She buried her daughter in a small communal plot. The heatwave was declared a public emergency. Officials praised the resilience of southern France. But resilience was a word

Camille had grown to hate. She wanted air that moved. She wanted a window that opened onto shade. She wanted a city that treated her pregnancy as something to be protected, not endured.

It was the third day without clean water in Jackson. Tameka leaned against the sink, trying not to panic as the water sputtered and stopped again. Hurricane had passed, but the damage to the treatment plant had left the taps dry and the toilets useless. She was 29 weeks pregnant and diabetic, instructed by her doctor to rest and monitor her blood sugar.

But how could she rest when there was no food in the fridge to eat and no water to flush waste?

Months ago, the local maternity ward had closed down. The next hospital was forty miles away, and her cousin's car was out of gas. Dre, Tameka's partner, tried calling the emergency line, but it went straight to voicemail. On the fifth day, the cramping began. Then came the bleeding.

She felt herself slipping into a fever. It wasn't the storm that did it. It was everything that followed — the silence, the neglect, the empty roads and broken promises.

She made it to the hospital only after a neighbour found a retired EMT willing to drive. Her son was delivered stillborn. With sepsis, Tameka spent two days in the ICU. She woke to a nurse patting her hand gently, saying, "You're lucky to be alive."

Lucky. She thought of the weeks without response. The months of boil-water notices. The state budget had passed without fixing the pipes. After giving her child the name Elijah, she buried him in the spot where her grandmother used to pray.

She later testified before the state legislature. Her voice trembled, but she did not stop. She told them what they already knew: that this was not a crisis of nature. It was a collapse of care. And it had a body count.

What happened to Yasmin and María, to Camille and Tameka, was not isolated. It was a sequence. A pattern repeating with local inflexions. Salt in the womb, smoke in the lungs, heat in the walls, silence in the pipes. Climate change does not begin in storms. It begins in systems — failing ones. And those systems speak first through the bodies of women.

"*Climate change amplifies existing health vulnerabilities — but nowhere is this more intimate, more deadly, than in maternal health. We are now seeing heatstroke in pregnant women in Europe, toxic smoke-linked miscarriages in South America, and salt-driven preeclampsia in South Asia. These are not regional anomalies. They are planetary signals.*"

— Dr Tolu Oni, Public Health Physician, C Urban Epidemiologist, University of Cambridge; Lancet Countdown Contributor (2023).

These stories are not footnotes to the Sahel. They are confirmations. We thought the epicentre had a single name, but the fault lines run global. The crisis spreads not by wind, but by indifference. And when the earth heats too fast, when the air thickens with violence, the womb becomes the first to tell the truth.

The question we left hanging at the end of the last chapter now sharpens. What happens when slow collapse meets sudden trauma? What happens when a body already weakened by salt and smoke and silence is struck by a storm that does not die?

We find our answer in Malawi, where the river has already begun to rise. We have traced the salt lines on Yasmin's swollen wrists, followed the soot trails into María's lungs, listened to the silence where Camille's child should have stirred, and stood beside Tameka as clean water failed to come. In four countries, on four continents, we watched climate violence enter the womb not as an abstract threat but as a measurable collapse — of breath, of blood pressure, of systems that were never built to protect them.

These women do not only show us the scale of the crisis — they show us its acceleration.

The conditions that once took generations to degrade a landscape now do so in seasons. What once simmered now scorches. Droughts no longer pause to let the soil recover. Fires return before the embers cool. And heat, no longer episodic, settles in like a verdict.

The frontline, now a spiral, which was once a line but is now tightening, quickening, indiscriminate. And there is perhaps no place on Earth where that spiral spins with more velocity and cruelty than in the Sahel. There, the heat no longer retreats. The rains, when they come, arrive as fury. It is in that fragile basin — scarred by fire, thirst, and forced migration — that we now confront the most terrifying symptom yet.

A cyclone with no end.

And if this is a global illness, then its most urgent signal will not come from the margins — but from the eye of the storm, where the disasters once called "natural" now have addresses, names, and due dates.

REFLECTION: THE SLOW GLOBAL CRISIS

To stand before Victoria Falls in the late afternoon is to witness crisis in its most acute form — loud, sudden, and impossible to ignore. The journey we have taken through the Sahel reveals the opposite: a crisis of attrition. One that does not announce itself with thunder, but erodes like wind over stone. It is the difference between a lightning strike and a terminal sigh. And yet, the further we've travelled, the clearer it becomes: the sigh is not confined to one region. The slow violence we've traced in the Sahel is no longer just a warning — it is a pattern, now surfacing in river deltas, rainforests, and concrete suburbs across the world.

This section began as a regional diagnosis — a four-stage collapse of a social-ecological body. In Chapter 4, we watched Lake Chad wither into ecological organ failure. In Chapter 5, we witnessed the autoimmune reaction, as societies consumed by scarcity turned inward.

Chapter 6 followed the haemorrhage: the forced migration of mothers and children in search of breath, bread, and mercy. Chapter 7 placed us in Neïma Patricia's hands, where emergency care stood in for the decades of global inaction that had made her tent necessary. Together, these chapters marked the descent of a body into shock.

But now, we must redraw the map. We have seen Yasmin in coastal Bangladesh, her blood pressure spiking from salt-laced water, a physiological echo of the Sahel's thirst. We have stood with María in the burning forests of Peru, her baby pulling air through smoke thick with the consequences of extraction. We have grieved with Camille in Marseille, where heat pooled in forgotten urban corners until it stilled the pulse of her unborn child. And we have walked beside Tameka in Jackson, Mississippi, as the storm passed, but the system failed, leaving her to labour without water, power, or rescue. The Sahel is no longer an epicentre. It is a mirror. A preview of a global condition accelerating in different dialects but always speaking through the same bodies.

This is the hidden cruelty of slow violence: it masquerades as background noise. A failed rainy season becomes routine. A preterm birth becomes unlucky. A maternal death becomes a private sorrow rather than a public indictment. And so, the collapse continues — not with a scream, but with a shrug. Entire communities fall through the cracks, not because we do not care, but because the data arrives slowly, in teaspoons of suffering spread across geography and time.

And this brings us to the most terrifying clarity of all. We have now examined a patient weakened by chronic dehydration, economic anaemia, infrastructural decay, and clinical neglect. What happens when that same body is struck not by erosion, but by impact? What happens when the long emergency of maternal climate stress meets a fast, explosive trauma that gives no time to brace, no room to respond?

This is no longer a hypothetical.

We have tracked the fever. We now face the seizure.

| PART III |

THE PERFECT STORM, AN ACUTE SHOCK AND SYSTEM COLLAPSE

We have examined a patient suffering from a chronic, wasting illness. Now, we must ask: what happens when that same, weakened body is struck by a massive, acute trauma? In this Part, we move from the slow violence of attrition to the "perfect storm" of system collapse. We will stand in the heart of a cyclone that refuses to die and chart its cascading failure — of infrastructure, of medicine, and of the very systems we thought would keep us safe. This is not a weather event. It is a time machine. It is the sudden, violent erasure of modernity.

The Kololo people named this place perfectly: Mosi-oa-Tunya, "The Smoke that Thunders." It is a name for an acute shock, for a crisis of immense and immediate energy. To stand here is to witness the physics of collapse in real time. Over a million litres of water per second, an entire river, simply vanishes into a chasm. It is a beautiful, terrifying, and contained catastrophe.

We are now entering an age of unnatural disasters, of man-made spectacles that mimic this terrifying power but are neither beautiful nor contained. The atmosphere, supercharged with the heat of our industrial fever, is no longer behaving according to the old rules. It is giving birth to monsters—cyclones that linger for weeks, relentlessly pummeling the land below. In our last reflection, we asked what happens when a chronically ill patient is struck by a massive trauma. This section is the answer.

We will move from the study of a system in gradual decline to the anatomy of a system in full cardiac arrest. This is the phenomenon of acute shock, where a single, overwhelming event triggers a catastrophic failure of all interdependent systems at once. A doctor knows this sequence well: the failure of one organ, like the heart, immediately places unbearable stress on the others — the kidneys, the lungs, the brain — leading to a rapid, cascading collapse. The same is true for a nation. The storm is the heart attack. The destruction of roads is the circulatory failure. The power outage is the neurological shutdown. The failure of water purification systems is kidney failure. And the resulting disease outbreak is the final, systemic infection.

Our journey will take us into the heart of one such event: Tropical Cyclone Freddy, one of the longest-lasting storms ever recorded. We will begin in the belly of the beast, exploring the frightening new meteorology that allowed it to exist. We will then stand on the ground in Malawi with Eliza, a mother enduring the private crisis of childbirth at the exact moment the public, planetary crisis tears her world apart. From there, we will chart the storm's terrible aftermath, the "triple crisis" of wind, water, and disease. Finally, amidst the mud, we will find the heroes fighting to save lives when every system has been washed away.

This is the new face of the climate emergency. It is not a slow, predictable decline. It is a sudden, violent deletion of the fragile systems of modernity that keep us alive.

It is the terrifying realisation that the wall between chaos and order is far leaner than we would imagine.

This is the anatomy of a collapse.

CHAPTER 9 | THE ENDLESS CYCLONE

In the midst of life's gentle rhythms, a storm was brewing – one that would shatter the tranquillity of the Nsanje District and forever change the lives of Eliza and Joseph.

We thought we had mapped the thresholds — drawn lines between seasons, named zones of risk, and coded every threat. But the storms are learning faster than our systems. What we once called a "natural disaster" now behaves more like a chronic illness — recurring, mutating, refusing to leave the body. And for Eliza, the diagnosis would not come from a satellite, but from the sound of the roof being torn from her mother's house.

In the fertile plains of southern Malawi's Nsanje District, near the broad, life-giving Shire River, the world is a symphony of green. It's late February, the height of the rainy season, and the air is thick with the sweet, loamy scent of wet earth. The maize in Joseph's compact field stands tall, its heavy ears promising a bountiful harvest. Eliza, his thirty-yearold wife, nine months pregnant with their first child, sits in the shade of a large mango tree, weaving a sleeping mat from dried reeds. Her movements are slow, deliberate, hampered by the immense globe of her belly where their daughter, Esita, rests.

That evening, Joseph came in from the field, his face beaming. "The best crop I have ever seen," he says, placing a hand on Eliza's belly.

"She brings us good fortune already." Their life is one of rooted, predictable rhythms. Their community is close; the women often gather by the river to wash clothes, their bright chitenje cloths a splash of colour against the brown water, their conversations a vital network of local news and mutual support. They are her safety net, a web of shared female knowledge. The rains this year have been unusually heavy, the river running higher than usual, but this is seen as a good omen. The world feels safe, ordered, and full of grace.

Half a world away, in the sterile, air-conditioned quiet of the Météo-France regional operations centre on the island of Réunion, there is no sense of peace. Here, a team of meteorologists keeps watch over a storm they have been tracking since it was a mere disturbance off the coast of Australia. Its name is Freddy.

As scientists, we are trained to see the world in systems. But what the meteorologists were seeing began to defy the known rules. Heat engines are the tropical cyclones, harnessing power from warm ocean water. Typically, they form, intensify, make landfall, and then rapidly dissipate. Freddy did not do this. After crossing the entire Indian Ocean, it slammed into Madagascar, but did not die. It re-emerged over the Mozambique Channel, reintensified, and hit Mozambique. And still, it did not die.

"The persistence is something we have never witnessed," we were later told by Dr Hélène Riviere, a lead forecaster. "It was behaving like a living thing, refusing to weaken. The storm was real; our understanding was what was incomplete."

The reason for Freddy's unnatural life force was clear on their sea-surface temperature maps. The Mozambique Channel was an anomalous pool of extreme heat, with water temperatures several degrees above average. It's the finest signature of global warming. Over 90 percent of the excess heat trapped by greenhouse gases has been absorbed by oceans, transforming the surface of the sea into high-octane fuel for cyclones.

Zebedee's work has focused on this terrifying new reality: we are creating storms that can feed not just on the ocean, but on the land itself. Under conditions of extreme soil saturation, the "brown ocean effect" can occur, where a storm draws latent heat and moisture from the flooded ground, maintaining its strength far inland. Freddy was a new species of storm, born of a heated planet.

The models on the screens at Réunion began to produce a forecast of unprecedented horror.

After seething over Mozambique, Freddy predicted a curve back for a second, even more powerful landfall, its path leading directly into southern Malawi, a landlocked region with no history of direct cyclonic winds. The scientists, trained in the dispassionate language of data, began speaking in hushed, worried tones. They were watching a slow-motion catastrophe unfold.

Returning to Eliza's village, the showers of the gentle afternoon do not stop. A steady, drumming rain that is no longer comforting continues through the entire night. By the next morning, the river has risen visibly, its colour changing to a churning, angry yellow. The low bank where the women did their washing is gone. Joseph spends the day on the roof of their house, adding extra ties of sisal rope to the thatch, his face grim.

"They are now consistently observing storms that behave in ways outside our historical records... These 'endless' storms are not just wind events; they are unprecedented rainfall events, capable of triggering inland flooding on a scale we have never had to model before."

— Dr Kevin Trenberth, Distinguished Scholar, National Centre for Atmospheric Research (NCAR)

The village chief, Chirwa, receives a visit from a flustered district official with a damp piece of paper. The warning is abstract: "Expect continuous heavy rains and abnormally strong winds." Chief Chirwa calls the elders. They have managed floods before. He sends young men through the village, telling everyone to secure their roofs and prepare to move to the schoolhouse on higher ground. The warning is being heeded, but the response is calibrated to a known threat. They are preparing for a flood. They have no concept of a cyclone.

In Réunion, the warnings have become frantic: "CATASTROPHIC," "LIFE-THREATENING," "EXTREME." They project rainfall totals that seem like typographical errors — over 500 millimeters in 48 hours. They know this will cause flooding on a biblical scale. They know the storm's slow pace means the winds will batter the same areas for days. How do you communicate a crisis that has no precedent?

The day before the storm hits, the rain is relentless. A profound, unnatural gloom has settled over the land. All the animals have fallen silent. Eliza and Joseph move their few precious belongings onto a raised platform. They will go to her mother's house tonight. That night, the wind begins to moan.

The final cross-cut is to the meteorological lab. Dr Riviere stands before the largest screen. It shows the final satellite image of Freddy as it makes its second landfall. It is a thing of terrible, perfect beauty, a monstrous, swirling vortex of immense power.

"God help them," she says. "They have no idea what's coming." For the scientists, it was a moment of professional dread.

For Eliza, huddling inside her mother's house as the first true gusts of wind began to tear at the roof, it was something far more primal.

The world outside stopped being wind and rain and became a single, deafening, physical force — a malevolent entity determined to tear their world apart.

The scientific abstraction was over.

The lived experience, in all its terror, was about to begin.

CHAPTER 10 | THE WALLS CAME DOWN

Collapse isn't always loud. Sometimes it arrives in the silence between a busy signal and a disconnected line. In the absence of transport. In a hospital that locked its doors years ago and never reopened. The erosion of a health system doesn't look like an explosion — it looks like a mother sitting on hold, bleeding, while the system decides if her life is worth the drive.

The world has shrunk to the four walls of her mother's home for her. Eliza had seen the last of the dreadful world outside before Joseph had closed the door. It was no longer their village. It was a grey, churning chaos. She saw her own neighbour's roof lift off in a single, impossible piece and fly into the gloom like a discarded leaf. That was the last thing she saw of the world she knew. Now, there is only the space inside.

The sound is a solid thing, a physical, relentless pressure that pushes on the mud-brick walls, on the thatch roof, on the very air in Eliza's lungs. It is a constant, deafening roar, punctuated by the sharp crack of flying debris striking the house. Eliza sits on a low stool, her back pressed against the central pole of the hut — the part her father always called the "heartwood." Her own heart hammers a frantic, terrified rhythm against her ribs. Her mother, Lina, moves with a strange, ritualistic calm, stuffing rags into the cracks around the doorframe, her lips moving in a silent, constant prayer. Joseph tests the main roof beam with the flat of his hand, as if he can feel the integrity of the wood, his face a mask of grim concentration.

The floor is now a chilly, slippery mud that sucks at their ankles, and a dark stain of water has been leaking beneath the door for the past hour. As her mother taught her, Eliza makes an effort to pay attention to her breathing. The nose goes in, and the mouth goes out, at a steady, calm pace to prevent her from being overcome by panic. Her pregnancy is nine months along. Her baby, Esita, is a treasured cargo in a sinking ship, a heavy, substantial weight inside her.

She thinks of the small, soft baby mat she finished weaving just last week, now packed in a bag that feels a world away. She must be calm for the baby. She must be a fortress.

Then, a new violence begins, one that comes from deep within her own body, not from the outside.

Something strong and indisputable begins with a low, coiling stiffness in her back that abruptly intensified. Her muscles compress into a knot of iron with their own intelligence as the pain spreads across her belly. She lets out a quick, uncontrollable gasp. Her throat tightens with each breath. It is not a Braxton Hicks, not one of the "practice" contractions she has felt for weeks. This is different. This has teeth. This is an invasion. The pain lingers for a long, timeless minute, a minute in which the storm outside ceases to exist, and then slowly, gradually, recedes, leaving her breathless and shaking.

Her body has betrayed her. A voice screams inside her head, all the nights in her life, a voice of pure, disbelieving rage directed at her own flesh, you choose this one? The sheer biological injustice of it is a separate, sharper pain than the contraction itself. Her body has chosen this exact moment, this storm of wind and water, to begin the most vulnerable journey of her life.

Another wave of pain hits, this one is stronger and commanding. She lets out a faint moan.

Joseph is by her side in an instant, his face near hers, his hand on her shoulder. "Eliza? "What is it?" he yells above the din.

She looks at him, her eyes wide with a new, dual terror. "The baby, it's coming," she manages to say.

As her mother hears, her stoical piece shatters for a second, replaced by a flash of raw fear before she masters it again. "It is the storm demon," Lina whispers, her voice filled with a terrible, ancient awe. "It is an evil wind. It wants the baby." Joseph shakes his head, dismissing the old belief, his mind racing for a practical solution that does not exist. "When the wind dies down, I will carry you," he says, his voice desperate. "I will carry you to the clinic myself." But even as he says it, they all know it is a fantasy. The clinic might as well be on the moon.

We know now, with the cold clarity of science, what was happening to Eliza. The extreme stress of the storm, the terror, the vibration and the constant, deafening noise had flooded her system with cortisol, the body's primary stress hormone. This massive cortisol surge, in turn, can trigger a premature release of the hormones that initiate labour — prostaglandins and oxytocin. The storm was not just the setting for her labour; it was very likely its direct, physiological cause. The planetary crisis had reached inside the walls of her home and her body and had flipped a switch.

The house begins to die. It doesn't happen all at once, but in a series of agonising, sequential failures.

First, a section of the thatch roof directly above them is torn away with a sound like ripping cloth. A cold, muddy stream of water pours in, drenching them instantly. Through the new hole, Eliza can see the sky, a swirling, malevolent grey. Then comes the sound of the main support beam, the heartwood, cracking. It is not loud, but it is a sharp, definitive report, like a bone breaking, followed by a shower of dust and dried insects.

The west-facing wall, the one taking the brunt of the cyclone's fury, begins to bulge inward, as if the house is taking a final, painful breath. A network of wet, black cracks spreads across its surface-like veins. Then, with a wet, sucking sound, the wall ceases to be a wall. It dissolves, it sags, it gives way, a cascade of stalk and mud collapsing into the ascending water inside the hut. The small space that was the protection of their lives is gone, erased. They are suddenly outside, while still being inside.

"Get down!" Joseph screams, his voice almost lost. He tackles Eliza, pulling her and her mother into the far corner of the hut, the one still partially protected. He throws his body over.

Eliza's a human shield against the wind and the flying debris. The storm is in the room with them now, a living, breathing entity of chaos.

Another contraction rips through Eliza's body, and this time she screams, her cry swallowed by the immense roar of the cyclone. Her water breaks, a sudden, warm gush that is indistinguishable from the cold rain and mud now pooling around them. They are lying in the ruins of their home, a small, terrified knot of humanity. The shock gives way to a dawning, horrifying realisation of their situation. They are cold, they are soaked, and they are utterly, completely exposed.

Eliza lies in the mud, her mind a vortex of pain and terror. The contractions are coming faster now, one wave crashing into the next. She is no longer a person. She is only a body at war with itself, in a world at war with everything.

The clinical, detached language of another world feels like a cruel joke. There is no protocol here. There is no evacuation. There is only the rain, the mud and Joseph's body, a human shield trying to ward off the end of the world. He is shouting something in her ear, with a hoarse and desperate voice, but she cannot figure out the words.

Despair, cold and absolute, begins to settle in, a chill deeper than the rain. The cold mud was seeping through her clothes, the water engulfing her. She has a moment of strange, out-of-body clarity, seeing the absurdity of it all: a woman trying to give birth in the ruins of a collapsing house, in the heart of a horrifying storm that is eating the world. This is it. This is where she will die. This is where her baby, her

Esita, will die before ever drawing a breath. The world is too loud, the pain is too great. She feels herself slipping away, a strange, inviting numbness spreading through her appendages. It would be so easy to just let go, to close her eyes and let the storm take her.

OBSTETRIC EMERGENCY PROTOCOL: COMPOUND DISASTER EVENT

"*In a resource-limited setting, an obstetric emergency (e.g., obstructed labour, postpartum haemorrhage) compounded by a Level 4+ natural disaster presents a mortality risk increase of up to 400% for both mother and infant. The collapse of infrastructure (transport, electricity) and health facilities renders standard clinical interventions impossible. Survival becomes contingent on immediate, non-clinical evacuation and improvised shelter.*"

— Dr Isabella Rossi, Head of Maternal Health in Crisis Zones, Médecins Sans Frontières (MSF).

She is on the verge of surrendering, of letting the roar consume her, when Joseph shakes her, his grip on her shoulder firm, painful. "Eliza! Look! A light!"

She forces her heavy eyelids open. Through the darkness of the driving sheets of rain, she sees it. At first, she thinks it is lightning, but it does not flash. It is a piercing beam, steadily sweeping through the maelstrom, cutting across the debris. It is the headlight of a vehicle, impossibly, miraculously here, navigating a landscape that should be impassable.

A voice calls out, a shout that is almost ripped away by the wind, distorted and strange. "Is anyone there? We are here to help!"

Eliza looks at Joseph, his face illuminated for a brief second by the sweeping light. She sees not just terror, but a flicker of wild, desperate hope. The pain of another contraction builds, but this time, something has changed. The despair has been pierced by a single, fragile ray of light.

It is not yet a promise of survival. But for the first time in hours, there is the possibility of one.

CHAPTER 11 | THE ENDLESS CYCLONE

In a time of famine, seeds are power. But in a world where climate change has weaponised scarcity, even a seed becomes currency — owned, priced, patented. What was once shared across generations now vanishes into vaults beneath frozen mountains, while communities that birthed this knowledge watch their harvests wither under foreign contracts and empty promises.

The wind has passed. After two days of a relentless, deafening roar, the silence that descends upon southern Malawi is not one of peace, but suffocating. It is the silence of a world erased.

Dr Samuel Mponda stands in what used to be the main corridor of the Nsanje District.

Hospital, and the only sound he can hear is the soft, surreal lapping of murky water against the walls. He is wading through half a meter of murky, brown floodwater. The power has been out for thirty-six hours, and his world is a moving bubble of light from the beam of his headlamp, cutting through the humid darkness. The air reeks of the cloying smell of river mud, wet plaster, and something else — the sweet, sickly scent of decay.

The full impact of the storm's fury is now starkly apparent. Through the beam of his light, he sees the scale of the destruction. The maternity ward, a section he had been particularly proud of, lies in ruins, its roof torn apart and tangled in a mess of twisted metal and waterlogged debris. The pharmacy, once a repository of life-saving medicines, is now a flooded nightmare, its contents a useless, floating sludge. The hospital's two ageing ambulances are sitting in the flooded car park, their engines and electronics drowned, their wheels invisible beneath the brown water. The hospital, once a bastion of order and healing for a population of a quarter of a million people, is no longer a hospital. It is a ruin.

He is not just a doctor anymore. He stands as a general surveying the aftermath of a war, his medical team a weary but resilient force. The cyclone may have won the battle, but they're determined to win the war. His staff, exhausted but running on pure adrenaline, are doing what they can. They have moved the few dozen patients they had before the storm to the only remaining structurally sound part of the building: the administration wing. They are working with the few supplies they could salvage. But the storm's primary assault, the acute shock of the wind and rain, is over. The true disaster, Dr Mponda knows, is just beginning.

A nurse, a young woman named Mercy whose face is pale with fatigue, sloshes through the water towards him. "Doctor," she says, her voice strained. "The boat from the Catholic mission just arrived. They brought six people they found on the roof of a trading post. Two of them… Doctor, they are very sick. Diarrhea. Like rice water. They cannot even hold down water." Even in the dim light, he can see the fear in her eyes; she knows what this means. He thinks of Eliza, the pregnant woman they had managed to rescue during the storm. They had gotten her here, delivered her baby girl safely by the light of a cell phone, but now the hospital itself had become a trap.

Dr Mponda feels a cold knot of dread form in the pit of his stomach. He has seen this before, years ago, after another flood. He knows that specific, clinical description. It is not just diarrhoea. It is a biological fire hose, a body attempting to expel itself from the inside out. He knows the name of the microbe that causes it, and he knows what it means for his ruined hospital and the thousands of displaced people huddled on any scrap of high ground they can find.

The third crisis has arrived. Its name is cholera.

The dread Dr Mponda felt was not just the fear of a single disease. It was the recognition of a classic, deadly pattern, a domino effect that turns a natural disaster into a full-blown humanitarian catastrophe. As a doctor and a scientist, I have studied this lethal sequence. It is the anatomy of a system collapse cascade.

The primary shock was the cyclone itself, an event of unprecedented scale and duration for this region. The secondary shock was the flood. As Zebedee's work has shown, the record-breaking rainfall from a slow-moving storm falling on already saturated ground created flash flooding that overwhelmed every natural and man-made drainage system. The great Shire River, normally a source of life, burst its banks and became a destructive force, swallowing villages whole.

This flood, in turn, is the perfect delivery mechanism for the tertiary shock: the disease outbreak. Floodwater transforms the landscape into a breeding ground for pathogens. The floodwaters inundate thousands of pit latrines, capturing their contents and mixing them into a vast, swirling soup of human waste. This toxic water then infiltrates wells and boreholes, contaminating the only sources of drinking water. The entire environment becomes a transmission vector for waterborne diseases. And the most efficient and terrifying of these is Vibrio cholerae.

The cholera bacterium is a masterpiece of evolutionary engineering. When ingested, it takes hold of the small intestine and releases a potent toxin that essentially hijacks the body's cellular function, forcing intestinal cells to pump out massive quantities of water and electrolytes. The body's desperate attempts to purge the infection result in brutal, watery diarrhea, with severe cases losing up to a liter of fluid hourly. Their blood thickens to sludge.

Their organs, starved of fluid and electrolytes, begin to shut down. Death, from profound dehydration and hypovolemic shock, can occur in a matter of hours.

For a pregnant woman, the diagnosis is a nightmare squared. Dr Robinah knows this fight intimately. A doctor treating a pregnant woman with cholera is battling two wars at once. The severe dehydration can dramatically reduce blood flow to the placenta, starving the fetus of oxygen and potentially causing a catastrophic placental abruption. The violent intestinal cramping and hormonal chaos can easily trigger pre-term labour. All the while, the mother's own body is sliding towards circulatory collapse. It is one of the most complex and dangerous balancing acts in all of medicine: trying to save the mother from dying of dehydration, while simultaneously trying to prevent her body from sacrificing the baby in the process.

We can trace the journey of the single bacterium that will soon place Dr Mponda before an impossible choice. It begins in a flooded latrine twenty kilometres upstream from his hospital. It is washed into the main current of the Shire River. It travels downstream for hours, part of a vast, invisible microbial tide. It eventually swirls into a flooded tributary and contaminates a shallow well on a small hill where a hundred people, their homes washed away, have taken refuge. A mother, desperate to soothe her thirsty child, lowers a bucket into this well. Later that day, her child develops diarrhoea. Two days later, a rescue boat finds the small, isolated community and evacuates the sickest among them, including this mother and her now critically ill child, to the nearest functioning facility: Dr Mponda's ruined hospital. The microscopic invader has just breached the last line of defence.

Back in the makeshift ward of the Nsanje District Hospital, the situation has become desperate. Dr Mponda now has over two dozen patients showing the clear signs of cholera. They lie on mats on the cold, damp floor, their bodies wracked with cramps. The air is thick with the smell of sickness and disinfectant. The primary treatment for cholera is simple: rehydration. Pumping the body full of fluids and electrolytes faster than the disease can expel them. For those who can still drink, there are oral rehydration salts. But for the most severe cases, the ones already sliding into shock, the only treatment is intravenous fluids — a direct line of saline solution into the veins.

"*Major foodborne and cholera outbreaks are closely linked, especially in areas with poor water and sanitation infrastructure. A flood not only destroys homes, but it also contaminates the water system, affecting the population. In the context of climate change, every food warning must also be treated as a public health and infectious disease emergency.*"

— Dr Dominique Legros, Cholera Program Lead, World Health Organisation.

Dr Mponda is down to his last few bags of saline. The main pharmacy is underwater, and all roads to the capital are impassable. There will be no resupply for days, perhaps weeks.

He is kneeling between two patients. One is a little girl, no older than five, the child from the upstream village. Her eyes are sunken, her breathing shallow, and her skin has lost elasticity. She is in the final stages of dehydration. The other patient is a woman in her third trimester of pregnancy, brought in from another rescue boat. Her body is trembling with violent cramps, both from the cholera and from what are clearly pre-term labour contractions. Both need IV fluids immediately to have any chance of survival.

Mercy, the nurse, appears at his side, her face grim in the bouncing light of his headlamp. She is holding up a single, clear plastic bag of saline solution. “Doctor,” she says, her voice cracking. “This is the last one.”

This is the endpoint of a system collapse. It is a moment of terrible, quiet intimacy where all the grand failures — of policy, of infrastructure, of preparation — are distilled into one doctor's impossible choice between one life and another. He looks at the small, still form of the child.

He looks at the pregnant woman, her hand clutching her belly, fighting for two lives at once. The last bag of saline solution in his hand feels impossibly heavy, the weight of a world that has broken its promise to them both.

When the hospitals are gone, when the medicines are ruined, when the doctors are overwhelmed, the very definition of 'care' must change. It must retreat from the buildings and the institutions. It must find a new home in the mud, in the tents, in the skilled, steady hands of those who refuse to surrender.

It must become midwifery.

CHAPTER 12 | THE ENDLESS CYCLONE

In the age of climate collapse, war does not wear a uniform. It wears the mask of a development deal, a conservation easement, a security alliance. Where rainfall disappears, alliances shift, and where forests are fenced, fighters emerge. The battlefield is no longer defined by ideology, but by access to life's last essentials: land, water, and the right to stay rooted.

The silence after the storm is a lie. The wind may have ceased its screaming, but the world is now filled with the sound of water. It is a constant, mournful gurgle, the sound of the Shire River occupying a space where it does not belong, a landscape of homes and fields now transformed into a vast, brown, un-navigable sea.

In the days following Cyclone Freddy's passage, what is left of the Nsanje District Hospital is a scene of quiet, shell-shocked horror. The moment of Dr Samuel Mponda's impossible choice over the last bag of saline has become a grim legend among the staff, a symbol of their complete systemic failure. Dr Mponda himself is a ghost, moving through the ruins with a haunted look in his eyes, his authority washed away with the medicines in his pharmacy.

It is Mercy, the young nurse who had stood beside him, who rallies the survivors. "Doctor," she says to him on the second morning, finding him staring at the collapsed roof of the maternity ward, "the people are looking to us. We cannot give up." Her voice is quiet, but it carries a steel that cuts through his despair. Before any outside help arrives, she begins to organise. She dispatches the local fishermen, whose small boats survived the storm, to check on outlying, submerged villages. She gathers the other surviving nurses and the village's traditional birth attendants — women whose knowledge she has always respected — and creates a census of the known pregnant and postpartum women. She helps organise a communal kitchen with the last of the dry grain. She is the first responder, because she understands a fundamental truth of disasters: help does not come from the outside in; it radiates from the inside out.

On the third day, the outside world finally arrives. It does not arrive as a grand flotilla of rescue boats, but as a single, mud-caked, white Land Cruiser with a blue UN logo on its door. It has taken the emergency response team from the UNFPA two days to navigate the hundred kilometres from the city of Blantyre. For Mercy and the other staff, the sight of the vehicle is a moment of profound, almost painful relief. It is the first sign that they are no longer alone.

Mercy does not wait to be given instructions. She meets the team leader, a seasoned Kenyan logistician named Evans, and in a rapid-fire, ten-minute briefing, lays out the situation. She then takes them to the one building in the village that is still mostly intact: the primary school, built of brick on a small rise of high ground. "This," she declares, "will be our new hospital."

What follows is a testament to the incredible power of community resilience. It is a partnership. The UNFPA team brings supplies and technical expertise, but Mercy and her community provide the essential local knowledge and manpower. Men from the village clear rubble and reinforce the damaged roof with plastic sheeting. Women form teams to scrub the mud from the floors, creating a clean, if spartan, space. Mercy moves through the chaos with a calm authority, directing the setup of a triage area, a small cholera treatment corner, and, in the largest classroom, a makeshift maternity ward. This is not a hospital built of bricks and mortar; it is a sanctuary built of shared will.

The scene unfolding in Nsanje is a textbook example of a Level 3 Emergency Health Response. As we have seen in our work, the logistics of such a response are a science unto themselves. Zebedee has studied the immense challenges: establishing a supply chain over roads that no longer exist, deploying mobile water purification systems into a cholera-contaminated zone, and setting up temporary power grids. But the aid that arrives is not just clinical. Alongside the boxes of medical supplies, there are others marked "Dignity Kits." Dr Robinah knows these are as essential as any medicine. Inside each kit is a bar of soap, a new wrap-around chitenje cloth, reusable sanitary pads, a comb, and a small, high-pitched whistle. The first items restore hygiene and a sense of personal cleanliness — an essential part of feeling human. The whistle is for safety, a tool to summon help in the chaotic, often dangerous environment of a temporary shelter. The kits are a profound acknowledgment that survival is not just about keeping the body alive; it is about protecting a person's spirit, their sense of self-worth.

Furthermore, Mercy and her team are trained in more than just clinical skills. They are practitioners of what is known as "psychological first aid." When a new mother arrives, shaking and traumatized, Mercy will sit with her for ten minutes, even in the midst of the chaos. She will not probe for details of the horror. She will simply offer a cup of clean water, make eye contact, and listen if the woman wants to talk. She will reassure her: "You are in a safe place now. You and your baby are our only priority." It is a form of care that acknowledges the unseen wounds, the trauma that, if left unaddressed, can be as debilitating as any physical injury.

Late one afternoon, a new patient arrives, brought in on a makeshift stretcher. Her name is Diana. She is a young woman from an outlying village, and she gave birth two days ago, at home, at the absolute peak of the storm. We can only imagine the terror: the roof groaning, the walls shaking, and in the midst of this planetary violence, the intimate, overwhelming violence of labour. With only her terrified husband to help, she had delivered her son onto a plastic sheet as the floodwaters lapped at the doorway of their hut. They and their newborn have survived alone since then, surrounded by water.

Mercy brings them into the makeshift maternity ward. Diana is weak but stable. It is the baby who is in immediate danger. He is small, frighteningly quiet, and his skin is cold to the touch.

Mercy unwraps his blanket; his breathing is shallow; his movements are listless. He is not feeding. These are the classic, terrifying signs of neonatal hypothermia and the onset of sepsis. In a functioning hospital, he would be rushed to a NICU and placed in an incubator. Here, that is not an option.

This is the moment where skill transcends technology. Mercy knows precisely what to do. She turns to Diana. "We are going to keep your baby warm," she says, her voice a mixture of gentleness and firm command. "You are going to be his incubator."

She helps Diana sit up and unbutton her blouse. She then takes the tiny, cold infant and places him directly onto his mother's bare chest, belly-down, his head turned to the side. She covers them both with every blanket she can find, tucking them in tightly, creating a sealed pocket of shared warmth. This is Kangaroo Mother Care (KMC), a low-tech, high-impact intervention. The science is as elegant as it is profound. The mother's chest provides a stable, self-regulating source of heat. The sound of her heartbeat and the rhythm of her breathing help stabilise the infant's. The skin-to-skin contact triggers a release of oxytocin in both mother and baby, reducing stress, promoting bonding, and stimulating the hormones necessary for milk production. Mercy then shows Diana how to express a few precious drops of colostrum — the nutrient-rich first milk — onto her fingertip and touch it to the baby's lips. It is medicine, food, and warmth, all delivered through the simple, robust technology of the human body.

For hours, Diana sits there, a living incubator, a human life-support system. Mercy checks on them every thirty minutes. It is a quiet, intense drama, a silent battle fought in the corner of a ruined schoolhouse.

"Kangaroo Mother Care is one of the most powerful and effective interventions in all of neonatal medicine. It is not a lesser alternative to an incubator; in many cases, it is superior. It is a perfect example of a life-saving technology that is human, not mechanical."

— Dr Nabila Zaka, Senior Health Specialist, UNICEF.

Just after midnight, a new sound fills the room, cutting through the weary silence. It is a small, tentative, but healthy cry. Mercy rushes over. The baby, warm now against his mother's chest, has woken up and is actively, instinctively seeking the breast. Diana, her face awash with tears of relief, guides him. He latches on. The sound of his suckling is the loudest, most beautiful sound in the world. As the other women in the makeshift ward hear the cry, a ripple of shared joy passes through the room. They are all mothers. They all understand the magnitude of this small victory. It is their victory, too.

Later that night, before trying to get an hour of sleep on a mat in the corner, Mercy stands for a moment and looks at Diana and her peacefully sleeping son. She feels a wave of exhaustion so profound it makes her knees weak. But she also feels a fierce, quiet pride. This is why she does this work. Not for the praise, not for the pay, but for this. For this one, single, breathing life.

The sound of that healthy baby's cry in a flooded schoolhouse in Malawi is a testament to the incredible power of human ingenuity and courage. It is a moment of pure, hard-won hope. But it is a hope forged in the crucible of absolute system failure. It is the response, not the solution.

And it raises a haunting question. While this life-or-death drama unfolds on the climate front lines, what of the world that created the storm in the first place? Is the crisis truly contained to these distant epicentres of mud and heroism? Or is the same planetary fever that floods the fields of Malawi now silently, invisibly, beginning to heat the nurseries of the West?

REFLECTION: THE PERFECT STORM

To stand near Victoria Falls at night is an entirely different experience. The visual spectacle of the day vanishes into the deep darkness, and you are left with only the sound and the feeling. The roar becomes deeper, more elemental. The ground beneath your feet continues to vibrate with the ceaseless, indifferent power of a million litres of water per second plunging into the abyss. You are humbled not by what you can see, but by what you can feel: a raw, untamable, and permanent force.

The journey we have just completed through the heart of Cyclone Freddy was a journey into a similar kind of darkness, one where the familiar visuals of modern life were stripped away, leaving only the raw, elemental forces of nature and human survival. Part II of this book was the story of a slow, chronic disease. Part III was the story of an acute, massive trauma. And having witnessed it, we have come to a ground-breaking and deeply unsettling conclusion.

We have been using the wrong language to describe the threat of these new, supercharged storms. We speak of resilience in terms of infrastructure — of building higher sea walls, stronger houses, and better drainage systems. The story of Cyclone Freddy teaches us that this is a dangerous illusion. The ultimate threat of an acute climate shock is not the destruction of property.

It is a system collapse. It is the violent, high-speed erasure of the complex web of interconnected services — healthcare, sanitation, electricity, logistics, governance — that creates the illusion of safety we call 'modern life.'

The cyclone was not just a weather event; it was a time machine. In the space of forty-eight hours, it threw a 21st-century community back into a 19th-century reality. Consider Eliza's ordeal. Her harrowing birth was not just a personal crisis; it was a journey backwards through medical history. She was stripped, hour by hour, of every advancement of the last 150 years. Consider Dr Mponda's impossible choice. His dilemma over the last bag of saline was the logical endpoint of this systemic regression, a return to a pre-antibiotic, presupply-chain state of medicine. And consider Mercy's heroism. Her midwifery in the mud was not just an act of profound courage; it was the rebirth of the most ancient form of care precisely because every modern system had been annihilated. The success of a low-tech intervention like Kangaroo

Mother Care is both inspiring and a terrifying indictment. It proves that the most resilient technology we have is the human body and the knowledge of how to care for it. But it is a solution that should never have been necessary.

This is the ground-breaking lesson of the perfect storm. Climate change is a force of radical de- modernization. As storms intensify, our intricate systems are exposed as fragile and vulnerable. We have built a world that is exquisitely optimised for a climate that no longer exists. And the first casualty of that world's collapse is the assumption of safety.

It is tempting to view the story of Malawi's collapse as a tragedy of a distant and uniquely vulnerable place, to find comfort in the strength of our own infrastructure. But this is the final and most dangerous illusion. The physics of a supercharged atmosphere are universal. The fragility of our complex, interconnected systems is a feature, not a bug. The disaster that devastated Eliza's world was a localised event, but the forces that created it are global. What we witnessed in Malawi was not an exception. It was a dress rehearsal. And it forces us to ask a deeply unsettling question: are any of us truly safe?

| PART IV |

A SHARED PERIL — GLOBAL CONNECTIONS AND CULTURAL SHIFTS

For half a century, we have lived with an illusion: a firewall built of wealth and geography that separates "their" vulnerability from "our" safety. It is tempting to see the collapse in Malawi as a distant tragedy. But the firewall is a fiction. It is burning. In this Part, we trace the pathways of a shared peril, following the embers from the Sahel to the unspoken front lines in the West. We will find the same planetary fever in the air of a Californian suburb and see how it invades not just our bodies, but the very fabric of our cultures. There is no "us" and "them." The peril is shared because the air is shared.

As dusk falls over Victoria Falls, a remarkable transformation occurs. The visual spectacle of the day fades to the deep, resonant darkness of the African night, but the human landscape comes alive. Standing here at the precipice, we are surrounded by the languages of the world. We hear snippets of German, of Japanese, of Spanish, of English from Texas and London. People who live their lives continents apart, in vastly different cultures and economies, stand shoulder to shoulder, their faces illuminated by the glow of their phones, all trying to capture the same elemental truth of the thundering water. For a brief, powerful moment, the world feels small, interconnected, and bound by a shared sense of awe.

For half a century, the developed world has operated under the unspoken assumption that this sense of a shared planet did not apply to the consequences of climate change. There was an implicit belief in a firewall, a barrier built of wealth, technology, and geography that separated the "safe" world from the "vulnerable" one. This belief was a cornerstone of a late 20th-century economic model predicated on endless growth and the systematic offshoring of environmental and human costs. The droughts, the floods, the famines — these were seen as tragedies, yes, but as tragedies happening over there, on the other side of the firewall, to be witnessed on a screen with a sense of pity, but not of personal peril.

At the end of our last reflection, we asked if any of us is truly safe. We are now reporting that the firewall is a fiction. It is burning. The embers are already landing on our side of the wall.

In this next part of our journey, we will trace the pathways of this shared peril across two distinct but interconnected fronts.

The first front is physical. The same planetary fever that supercharged Cyclone Freddy and withered the fields of the Sahel is a global phenomenon. It does not respect borders. While it may manifest as a catastrophic flood in Malawi, it expresses itself as an unprecedented heat dome over Phoenix, or as a wildfire that chokes the air of a Californian suburb. We will travel from the front lines we have witnessed to the unspoken front lines in the very heart of the developed world, entering the air-conditioned delivery rooms and OB-GYN offices to find that the bodies of pregnant women in the West are also beginning to register the strain of a destabilised climate.

The second front is not physical, but cultural and psychological. A crisis of this magnitude does not just alter landscapes; it alters societies. It puts unbearable pressure on the ancient traditions that have guided communities for millennia, in some cases strengthening them, and in others, breaking them apart. But it also creates new, modern anxieties. It is forcing young people in wealthy nations to ask a question their parents never had to consider: Is it morally responsible to bring a child into a world that feels like it is coming undone?

We will explore this fraying rope of tradition and the birth of a new climate-driven culture. We will see how an environmental crisis becomes a crisis of faith, of family, and of the future itself.

There is no "us" and "them" in this story. There is only a single, interconnected human family, living in a single, shared home that is now running a fever. The peril is shared because the air is shared, the oceans are shared, and the fierce, protective love for our children is the most universal instinct of all.

Let us now turn our gaze homeward.

CHAPTER 13 | AN UNSPOKEN RISK IN A WESTERN CITY

Not all violence is visible. Some is systematised, dressed in compliance and passed down like an inheritance. In the world of aid, there are forms to be filled, protocols to be followed, and permissions to be granted — while a mother bleeds out waiting for someone to approve a budget line. The rules are rarely cruel by design. But cruelty doesn't need design — only obedience without imagination.

The jump cut is jarring. One moment, we are in the mud and heroic chaos of a Malawian field clinic where a midwife like Mercy fights for one life at a time. Next, we are plunged into the cool, sterile, and silent calm of a perinatal diagnostic suite at one of Northern.

California's leading hospitals. The air is filtered, scented with a faint, antiseptic cleanness. The only sounds are the gentle hum of the ultrasound machine and the soft, rhythmic whooshing of a fetal heartbeat, a sound engineered for maximum reassurance.

Claire and her husband, Ben, had been laughing in the waiting room just moments before. They were scrolling through paint swatches on his phone, and they were torn between "duck egg blue" and "sage green" for the nursery. At thirty-two, Claire is a picture of vibrant, conscientious motherhood. A graphic designer who works from their home in a leafy Marin County suburb, she has approached her first pregnancy with the same meticulous attention to detail she gives her professional projects. She eats organic, she practices prenatal yoga, she has read all the right books, and her prenatal vitamins are lined up by the kitchen sink. She has done everything in her power to create a perfect, protected environment for the child growing inside her.

The 20-week anatomy scan is a pivotal moment in modern pregnancy. For twenty minutes, a friendly technician has been guiding them through a tour of their developing son. As they gaze at the screen, they see the perfect, curved arch of his spine, the four chambers of his tiny, beating heart, the astonishing detail of ten fingers and ten toes. Claire feels a wave of profound, joyful love as she sees his profile on the screen. He is real.

But then, the technician's cheerful patter ceases. She begins taking the same measurement of the femur bone, over and over again, her brow furrowed in concentration. The silence in the room transforms from calm to tense. "I'm just going to have Dr Ramirez take a look," the technician says, her voice a little too bright. "She can explain the images much better than I can."

Dr Eva Ramirez is the kind of doctor one chooses for a high-risk pregnancy. She is sharp, compassionate, and a leading expert in maternal-fetal medicine. She enters the room with a calm smile, but her eyes are focused on the screen. She re-examines the images, her movements efficient and precise. After a long, quiet minute, she turns to them.

"Okay," she begins, her voice even. "The baby's anatomy looks perfect. Heart, brain, kidneys — all developing beautifully." Claire feels a rush of relief, but it is instantly checked by the doctor's serious expression. "However, his growth measurements are lagging. He's measuring in the 8th percentile for his gestational age. What this suggests is a condition called Intrauterine Growth Restriction, or IUGR. It means he isn't growing as fast as we'd expect." She points to a grainy, grey-scale image pulsing with colour. "And the blood flow from the placenta, which you can see here, appears to be slightly compromised. The placenta isn't delivering nutrients as efficiently as it should be."

Claire stares at her. The clinical, Latinate words feel like a foreign language. "But… I don't understand," she stammers, her mind racing, replaying every meal, every activity of the last five months, searching for the mistake she must have made. "I've done everything right. I eat well, I exercise, I don't drink, I don't smoke. How can this be happening?"

Claire's shock is understandable. She, like most of us, operates under the assumption that our bodies are private territories, influenced primarily by our personal choices. We believe that if we follow the rules, we can build a fortress of health around ourselves and our children. But new and terrifying evidence reveals that our bodies are not private fortresses. They are ecosystems, intimately and inescapably connected to the health of the planetary ecosystem that we all inhabit.

The walls of our homes, our cities, and even our own skin are far more permeable than we think.

To understand what is happening inside Claire's body, we must travel back in time one year, to the previous autumn. A massive "gigafire," born of years of drought and a week of extreme heat, had raged through the mountains north of her home. These are no longer normal fires; they are atmospheric events. They create their own weather systems, generating pyrocumulonimbus clouds that can inject smoke and aerosols into the stratosphere. For two weeks, a thick, acrid, yellow-brown haze had blanketed the entire Bay Area. The sun was a dim, orange disk in a sky the colour of rust. The smoke was so vast that it travelled on the jet stream, causing hazy sunsets in New York City. Staying indoors was not a sufficient defence against an environmental event on a continental scale.

Zebedee's work has focused on the chemistry of this smoke. It is not just ash. It is a toxic cocktail of carbon monoxide, nitrogen oxides, and, most insidiously, particulate matter less than 2.5 micrometres in diameter. This is PM2.5. These particles are so microscopically fine that they bypass the body's natural defences and pass directly through the lungs into the bloodstream.

Once in the bloodstream, as Dr Robinah's medical research can attest, these particles act as foreign invaders, triggering a low-grade, systemic inflammatory response throughout the entire body. It is similar to the body's reaction to a chronic infection. This inflammation is the ghost in the system. For a pregnant woman, it can be catastrophic. The placenta is an organ of incredible complexity, a delicate, tree-like structure of blood vessels where the mother's and baby's circulations meet. This organ is exquisitely sensitive to inflammation, which can cause microscopic damage to its blood vessels, creating tiny clots, reducing blood flow, and impairing its fundamental ability to deliver oxygen and nutrients. Emerging research now suggests the damage may be even more serious, with evidence linking PM2.5 exposure to increased risks of gestational diabetes and even to potential epigenetic changes, in which environmental stress can alter how a baby's genes are expressed. The environmental event is over, but the biological insult lingers.

Back in the consultation room, Dr Ramirez is gently guiding Claire through this new, terrifying landscape. She feels the weight of her responsibility. As a clinician, she is trained to deal in certainties, in evidence-based diagnoses. But she is practising medicine on the front edge of a new climate reality, where the research is struggling to keep pace with the lived experience of her patients. She has seen this pattern — healthy, low-risk mothers with unexplained placental issues — too many times to ignore it.

"Claire, I was looking at your chart," she says. "Were you here during the Kincade Fire last fall?"

"Yes," Claire says, confused. "It was awful. We were under an evacuation warning. The sky was orange. But… I wasn't pregnant then. What does that have to do with anything?"

And here, Dr Ramirez must become more than just a clinician. She must become a translator. "There's a growing body of research," she explains, her voice careful and precise, “that links significant exposure to wildfire smoke with adverse pregnancy outcomes, even when the exposure happens before conception. The hypothesis is that the inflammatory response triggered by inhaling those PM2.5 particles can have long-term effects on the vascular system, and that can sometimes affect the way the placenta develops and functions later on.” She can see the disbelief in their eyes. “We’re seeing similar inflammatory pathways and placental issues in women under extreme heat stress, like in that record-breaking heatwave in Phoenix last year. The trigger is different, but the body’s response is unnervingly similar”.

She pauses. “I cannot tell you with one hundred per cent certainty that the smoke is the cause of your baby’s growth restriction,” she continues. “The human body is too complex for that kind of certainty. But I can tell you that what we are seeing on your scan is consistent with a pattern that other researchers and I across the West Coast are beginning to see with alarming frequency.” She then shifts into a new mode, outlining the plan. "Given the situation, we'll consider this a high-risk pregnancy, we'll take a more vigilant approach and closely monitor your pregnancy to ensure the best possible outcome."That means weekly biophysical profiles and non-stress tests to check on the baby, and regular ultrasounds to monitor his growth. We will watch him very, very carefully.” The fire from last year is now dictating Claire’s entire medical reality for the next four months.

"We are just beginning to understand the profound reproductive health impacts of wildfire smoke. Our studies are showing statistically significant associations between high exposure periods and increased risks of pre-term birth, low birth weight, and hypertensive disorders of pregnancy. We are, in effect, running an uncontrolled experiment on an entire generation of pregnancies."

— Dr Amy Padula, Associate Professor, Department of Obstetrics, Gynaecology & Reproductive Sciences, University of California, San Francisco (UCSF) .

Claire and Ben drive home in a state of stunned silence. The world outside their car, the familiar, beautiful Northern California's landscape, now looks different. It looks menacing. Ben finally breaks the silence. "We'll get a second opinion. We'll find the best specialist in the state. We'll fix this."

Claire just shakes her head, a tear finally tracing a path down her cheek. "Fix what?" she says, her voice barely a whisper. "The air? Where would we even go, Ben? It's everywhere."

Her words hang in the car, a recognition of a terrible, new powerlessness. This is the unspoken risk made real. It is a new, creeping anxiety that is beginning to permeate the experience of parenthood in the developed world. It is a form of grief for the loss of a world that felt predictable and safe. It is the end of the modern myth of control, the illusion that we can protect our children simply by making the "right" personal choices in our own homes. The environment is no longer a passive backdrop to our lives; it is an active and often hostile participant in our reproductive health.

CHAPTER 14 | THE FRAYING ROPE OF TRADITION

The new frontier of climate justice is paved with carbon credits — and stained with erasure. A forest in Congo is valued more alive than the people who live within it. A project's success is measured in tons of avoided emissions, not in babies safely delivered. And somewhere in the fine print, a mother is told that her survival is worth less than a ton of carbon stored in a tree.

In a small, dust-swept village in the Somali Region of Ethiopia, a girl of fourteen is memorising a passage from her science textbook. Her name is Aisha. The book is a tattered, precious object, its pages soft with use. She sits just inside the doorway of her family's hut, using the last of the day's light. She loves the elegant certainty of the diagrams, the way they explain how a seed becomes a plant, how a cloud becomes rain. It is a world of logic, of cause and effect, a world that makes sense. Her dream, a secret she holds close to her heart, is to become a teacher.

Her father, a kind, quiet man named Yusuf, approaches and sits before her. She sees the shame in his eyes before he speaks, and her own heart begins to beat with a slow, heavy dread. Yusuf was once a proud pastoralist with a herd of over fifty goats. The multi-year drought has taken them all. He was the one who first enrolled Aisha in the small village school, his chest swelling with pride as she learned to read and write. "You will have a different life from mine," he had told her. "Your mind will be your herd." Now, that promise has turned to ash in his mouth.

He explains it in a low, broken voice. An older man, a trader from a neighbouring village who still has some wealth, has asked for her hand in marriage. He has offered a bride price: four bags of grain and a single, bony, but living cow. It is a fortune. It is enough to keep her mother and her two younger siblings alive for the next six months. The trader's family arrives the next day to finalise the arrangement. The conversation is muted, sombre. It does not feel like a celebration. It feels like a funeral. Aisha watches from the hut as the man who will be her husband inspects the cow, a gesture that seals her fate.

She looks from her father's face, etched with a grief that mirrors her own, to the science textbook in her lap. She looks at the diagram of the water cycle — the sun, the evaporation, the clouds, the rain — a perfect, benevolent circle that has been broken. The orderly world of her book has been shattered by the disorderly, merciless reality of the empty sky. The promise of her education is being erased. This is not a cruel choice made by her father; it is a choice that has been forced upon him by a sky that has refused to rain. The transaction is not for a daughter; it is for survival.

Aisha's story is not a story of a cruel tradition. It is a story of a compassionate tradition — the protection of a family through a daughter's dowry — being twisted into a desperate, harmful last resort by a force it was never designed to withstand. As we have travelled through these communities on the brink, we have seen how the climate crisis is fraying the very rope of tradition that has allowed them to survive for centuries. This rope, as we have come to understand it, is woven from three essential strands.

The first strand is economic stability, provided by a predictable relationship with the land and the weather. The second is social cohesion, the dense network of family and community that provides a safety net of mutual obligation. The third is traditional ecological knowledge, the library of wisdom about how to live within the specific rhythms of a place. Climate change, driven by an extractive global economy that has externalised its costs onto communities like Aisha's for centuries, is not attacking one of these strands; it is a corrosive agent attacking all three at once, causing the entire rope to unravel.

The perversion of marriage customs is a direct result of the first strand snapping.

Zebedee's analysis of the data from UNICEF confirms this: in regions of the Sahel and Horn of Africa that have experienced more than two consecutive failed rainy seasons, rates of child marriage have, in some cases, doubled. For Dr Robinah, this is a medical catastrophe. A girl of fourteen is not a woman. Her body is not ready for childbirth. Her pelvis is still growing, often too narrow to allow a baby's head to pass safely, leading to a dramatically increased risk of obstructed labour. This, in turn, can lead to obstetric fistula — a devastating, life-altering injury that causes lifelong incontinence and social ostracism.

The second strand, social cohesion, frays when people are displaced. We met a young mother named Bisa in the Kalambari camp. She had a healthy baby boy, thanks to the intervention of Neïma Patricia. But in the weeks following the birth, a deep, listless depression had settled over her. One afternoon, we saw her struggling, near tears, as her baby cried, refusing to breastfeed. In her village, an experienced auntie or her own mother would have been at her side in an instant, gently correcting the baby's latch, offering a word of encouragement, and taking the child for a while so Bisa could rest. This traditional postpartum "social cocoon" is a critical component of maternal health. But Bisa's village was gone, likely destroyed in a conflict over a well, just like the one Oumar fled. Here, in the sprawling, anonymous chaos of Kalambari, she was utterly alone. Her shame and frustration grew, creating a cycle of stress that further inhibited her milk production. This isolation, as Dr Robinah knows, is a direct risk factor for severe postpartum depression, which can affect not just the mother's well-being, but the baby's development and their crucial bond.

But tradition is not merely a fragile thing that breaks. It can also be a source of profound resilience, capable of adapting. We witnessed this in Kalambari. We spent an afternoon with a group of grand-mères, the respected female elders of one section of the camp. They were the ones who, back in their villages, had most staunchly defended the tradition of large families as a sign of God's blessing.

We watched as one of them, Gogo Adama, held an informal meeting with a group of young, newly married women. At first, another elder in the group was resistant. "This talk of spacing children," the other woman said, her voice sharp, "it is the white man's thinking. Our strength has always been in our numbers."

Gogo Adama listened patiently, then responded, her voice firm. "Our strength was in our numbers when the land could feed our numbers. Here, what work is there for ten children? To have many children you cannot feed is not a sign of wealth; it is a sign of suffering." She then used a traditional proverb about planting seeds. "If you plant the maize too close," she said, "the stalks are weak, and the harvest is nothing. A good farmer gives each seed its own space to grow strong. It is the same with our children. Our wombs are our fields now. We must be good farmers."

"*The most successful public health interventions are those that are championed from within a community. When a respected elder, using the language and metaphors of her own culture, advocates for a new practice like family planning, the impact is a thousand times more powerful than any poster or pamphlet from an outside agency.*"

— Dr Fatou Sow, Senegalese Sociologist and Gender Specialist.

In this moment, Gogo Adama was performing an act of radical cultural adaptation. She was not abandoning her role as a keeper of wisdom; she was adapting that wisdom to a new, harsh reality. It was a stunning display of leadership, showing that tradition can be a source of innovation, not just a barrier to it. This adaptation, a shift in social norms, is a form of climate resilience as powerful as any seawall or drought-resistant seed.

The third strand of the rope — traditional ecological knowledge — is perhaps the most tragically frayed. When a community is displaced, a library is burned. Generations of knowledge about which plants are edible during a drought, how to read the behaviour of insects to predict the rains, which grasses best hold the soil, all of it is lost. This is an irreplaceable loss not just for the community, but for all of humanity, as we search for ways to adapt to a changing planet.

The fraying of these ancient ropes of tradition reveals the profound, often invisible, depth of the climate crisis. But the sight of a village elder advocating for family planning reveals something else: a spark of adaptive genius, a refusal to surrender.

The stories of this chapter — of both devastating loss and incredible resilience — force a final, critical question. If we know the risks, and we can see both the destructive and the creative ways humanity is responding, what then must be done? It is not enough to simply chart the pathology and witness the courage of the survivors.

We must now turn our full attention to the fight for the future, to the innovators, the advocates, and the architects of a more resilient world.

REFLECTION: A SHARED PERIL

It is twilight on the Victoria Falls Bridge. Below us, the Zambezi River thunders into its chasm, a violent, indifferent border between Zimbabwe and Zambia. To stand here is to be suspended between two nations, two histories, two realities, connected by a single, remarkable feat of human engineering. For most of modern history, this is how the developed world has viewed its relationship with the climate crisis: as a distant spectacle, a thunderous problem happening on the other side of a vast chasm, a chasm we believed was bridged only by our pity or our aid.

The journey we have just completed in Part IV was a journey across that bridge, but the destination was startling. We found that the peril is not on one side of the chasm or the other. It is in the very air that connects us. We have come to understand that the firewall we believed was protecting us is a myth. The defining characteristic of the climate crisis is not its distance, but its terrifying, radical intimacy.

We have been using the wrong metaphors. We thought of climate change as an external threat — a storm gathering on the horizon. But the stories of Claire in California and Aisha in Ethiopia reveal a more profound truth. The threat is not just external. It is invasive. It has breached the final, most sacred firewalls: the walls of our homes, the walls of our cultures, and the walls of our own bodies.

Consider the pathways of this invasion. For Claire, in her pristine Californian hospital, the invasion was microscopic and biological. Particulate matter from a distant wildfire, a symptom of the planet's fever, crossed from the atmosphere into her lungs, into her bloodstream, and ultimately, across the placental barrier. A crisis that began in the sky became a cellular crisis in the delicate, life-giving architecture of her womb. The peril became intimate.

For Aisha, in her drought-stricken Ethiopian village, the invasion was economic and cultural. The same planetary fever, manifesting as a lack of rain, crossed from the atmosphere into the soil, into her family's herd, and into the foundation of their economic survival. From there, it invaded the sacred space of her family's home, forcing her father into an impossible choice that extinguished her future. A crisis that began in the sky became a heartbreaking transaction that unravelled the fabric of her life.

One is a story of high-tech medicine; the other, of ancient tradition. But they are the same story. They are the story of how a vast, impersonal atmospheric phenomenon becomes a life-altering, deeply personal event. This is the ground-breaking realisation of our shared peril: climate change dissolves the boundaries we thought were solid. The boundary between the environment and the body, between public air quality and private reproductive health, between a global weather pattern and a father's decision about his daughter — all of it dissolves. The atmosphere is not just around us; it is inside us, shaping our biology and our choices. The womb is not a fortress; it is an ecosystem, exquisitely sensitive to the health of the larger ecosystem it inhabits.

To recognise the intimate nature of this crisis is terrifying. It can feel paralysing to accept that the very air we breathe can carry a hidden threat, that our modern lives offer no true sanctuary.

But it is also, paradoxically, where we find our power.

For if the crisis can reach into our homes, our families, and our bodies, then our response can also begin in our homes, our families, and our bodies. If the peril is shared, then so is the responsibility, and so is the courage.

The diagnosis has been made. The patient's condition, from the acute, crashing trauma of a super-storm to the slow, chronic wasting of a society, is now understood in all its terrifying complexity. The time for observation is over. The time for treatment, for intervention, for a fierce and intelligent resistance, has come.

It is time to join the fight for the future.

| PART V |

THE FIGHT FOR THE FUTURE

After such a harrowing diagnosis, despair is a logical response. But it is a luxury the people we've met cannot afford, and a conclusion the evidence does not support. The purpose of a diagnosis is not to pronounce a death sentence. It is to create a treatment plan. In this final part, we move from pathology to response. We will meet the innovators, advocates, and healers who are fighting for the future. We will examine the "system shock" of the recent past to understand what true resilience requires. And we will propose a new way to measure what truly matters. This is the story of that fight, an invitation to move from witness to participant.

Night falls quickly at the Zambezi's edge. The vibrant colours of the day collapse into a deep, profound darkness, and the world is remade. The visual spectacle of the Falls is gone, but the roar of the water continues its eternal sermon, and above, the vast, clear African sky reveals a canopy of stars with an astonishing, brilliant clarity. It is often in the deepest darkness that you can see the farthest.

After the journey we have taken together — through the fevered wards, the flooded plains, and the fire-scarred landscapes — it would be easy to succumb to the darkness. It would be easy to conclude that the forces we have witnessed are too vast, the systems too broken, and the suffering too immense to overcome. Despair is a logical, perhaps even a rational, response.

But that is a conclusion born of distance. It is a luxury the people on the front lines of this crisis cannot afford. And more importantly, it is a conclusion that the evidence does not support.

On the ground, amidst the ruins and the refugee camps, in the quiet of the labouratories and the noisy halls of policy, we found something else entirely. We found a fierce, intelligent, and unrelenting fight for the future. We found people who looked directly into the heart of the crisis and, instead of turning away, chose to act. They are the architects of resilience, the healers of a wounded planet, the defenders of the next generation. This final part of our book is dedicated to them.

Our journey from here is no longer a descent into the pathology of the crisis, but an ascent into the landscape of the response. We will meet the innovators — the scientists, engineers, and community health workers who are pioneering brilliant, often low-cost, solutions to protect mothers and babies from the immediate threats of a heating world. We will meet the advocates and policymakers who are fighting to correct the "glaring omission," working to embed the health of women and children into the very centre of our global climate strategy. We will argue that the inertia of our existing political and economic systems is not an immovable object, but a series of choices that can be unmade and remade.

The hope you will find in these pages is not a soft, easy optimism. It is not a promise that the fight will be simple or that victory is assured. It is a hard-won, pragmatic hope, forged in the crucible of the crisis itself. It is the hope you find in the steady, skilful hands of a midwife like Neïma. It is the hope you find in a scientist's brilliant, world-changing insight. It is the hope you find in a community's stubborn refusal to be erased.

This is the story of that fight. It is an invitation to move from witness to participant, to understand that the antidote to despair is not hope, but action. It is a roadmap to finding your own place in this, the most essential struggle of our time.

Let us meet the people who are fighting for the future.

CHAPTER 15 | THE ADAPTATION GAP

Recovery isn't a return. It's a reimagining. In a post-crisis world, where nothing will ever be the same again, rebuilding means inventing structures that never existed before — networks of care where there were only gaps, systems rooted in dignity where only survival was offered.

Rebuilding is not what governments do after disasters. It's what women do every single day, with what little is left.

As we navigate the complexities of a planet in crisis, one question looms large: Is there a cure? A single, magical solution might be appealing, but the reality is that addressing this vast and multifaceted challenge requires a diverse array of approaches. Yet, in the midst of this crisis, we have discovered beacons of hope. The front lines of the climate crisis are not just zones of suffering; they are also, astonishingly, vibrant laboratories of hope, teeming with a fierce and practical ingenuity.

The distance between having the tools and knowing a risk to mitigate it is what experts call the "adaptation gap." It is the space between a diagnosis and a treatment. For the pregnant women we have met in this book — for Fatima in her hut, for Zara on her long walk, for Eliza in the storm — this gap can be a death sentence. The innovators we will now introduce are the engineers of survival, the architects of resilience. They are the people working tirelessly to close that gap, not with grand, multi-billion-dollar mega-projects, but with brilliant, targeted, and often beautifully simple solutions. They are the Qs to this story's Bond, forging the tools for the fight ahead.

Our first stop is a sun-scorched village in the state of Rajasthan, India, a region that now regularly endures heatwaves that push the limits of human survivability. Here, we meet Priya, a young, relentlessly energetic engineer from Delhi. She is not in a pristine lab, but in the cramped, stiflingly hot courtyard of a small, concrete-block home, kneeling beside a pregnant woman named Sunita. Between them is Priya's invention: a simple, box-shaped device made of reddish-brown terracotta.

It's called the "Cool-Cradle." It is a non-electric cooling unit based on the ancient principle of evaporative cooling, the same phenomenon that makes you feel cool when you sweat. The device consists of a nested set of porous terracotta walls with sand in between. In a small, humble abode, a revolutionary cooling system brings life-saving relief. A simple, solar-powered device, crafted from local materials for under \$30, uses evaporation to chill the air. A tiny fan, powered by a small solar panel, circulates a gentle breeze, reducing the temperature by 8 degrees Celsius (15°F). For Sunita, eight months pregnant and struggling with heat exhaustion, the cool air is a godsend. As she feels the gentle caress of the cooled breeze, her face relaxes, and relief washes over her. This isn't just a device – it's a lifeline.

Thousands of miles away, in the mountains, we meet another innovator in the rural north of Rwanda. Jean-Pierre, a biomedical technician with a quiet, intense focus. His enemy is not heat, but time. The leading cause of maternal death in this region is postpartum haemorrhage — bleeding after childbirth. It is a condition that is almost always survivable with a timely blood transfusion. But for a woman who gives birth in a remote, off-grid clinic, the nearest blood bank can be a six-hour journey over treacherous mountain roads. It's often too late, as she gets there. Blood requires a reliable, constant refrigeration, a luxury that the national power grid, prone to frequent outages, cannot guarantee.

Jean-Pierre's solution is parked outside the small clinic: a "Solar Blood Bank." It is a hyper-efficient, medical-grade refrigerator, insulated like a fortress, connected to a robust battery storage system and powered by a set of six solar panels mounted on the clinic's roof. Inside, neatly labelled bags of blood — typed, screened, and safe — are stored at a perfect, constant 4°C. The local midwife, a woman named Elodie, touches the cold glass of the refrigerator door with a reverence usually reserved for a religious icon. "Last year," she tells us, her voice thick with emotion, "we lost a mother of four. She was a strong woman. She bled after the birth. We did everything we could, but we had no blood. We watched her die. With this," she taps the glass again, "she would be alive today. Her children would have a mother."

Priya's Cool-Cradle and Jean-Pierre's solar fridge are more than just clever gadgets. They represent a fundamental shift in the philosophy of climate adaptation, a move away from massive, centralised projects and towards what Zebedee's work identifies as the cornerstones of true resilience. These solutions are decentralised; they do not rely on a national power grid or a complex supply chain that can be severed in a storm or flood. They are locally maintainable; a cracked terracotta pot can be replaced by a village potter, and a solar panel can be cleaned by anyone. And they are low-cost and scalable; they are designed to be affordable and replicable by the thousands. They are antidotes to the brittleness of the complex, just-in-time systems we saw collapse in Malawi.

The medical impact of these innovations, as Dr Robinah can attest, is staggering. The 8-degree drop in temperature from Priya's device is enough to prevent a pregnant woman's core body temperature from rising into the danger zone, warding off the inflammatory cascade that leads to heatstroke and pre-eclampsia. The single unit of O-negative blood in Jean-Pierre's solar-powered fridge can restore the circulatory volume of a haemorrhaging mother, pulling her back from the brink of hypovolemic shock and death. These are not incremental improvements; they are life-saving transformations.

But the adaptation gap is not just about physical technology. It is also about information. In a sprawling, informal settlement on the outskirts of Nairobi, Kenya, we meet a team of young tech entrepreneurs. Their office is a chaotic, vibrant room filled with laptops and whiteboards. They have developed a mobile application called Afya Mama, which means "Healthy Mother" in Swahili.

The app is a work of genius in its simplicity. It cross-references meteorological data from weather satellites with the GPS coordinates of registered pregnant women and community health workers. When the forecast predicts a high-heat event for a specific neighbourhood, the system automatically sends out a targeted warning via SMS — crucially, a simple text message that works on the most basic of mobile phones, not just a smartphone app. The messages are clear, actionable, and delivered in the local language: "HIGH HEAT WARNING for your area tomorrow. Drink three extra cups of clean water. Rest during the hottest part of the day. If you feel dizzy, tell your health worker immediately." One of the software engineers mentions that she was inspired by a report she read about the devastating, unaddressed risks of heat exposure for pregnant women in places like Phoenix, a problem her own grandmother in rural Kenya also faced.

Their pilot program has been successful, showing a measurable decrease in heat-related ER visits among the pregnant women in their cohort. But now, they face the true adaptation gap. They have a proven, life-saving tool. But to scale it up from a pilot of five thousand women to a national program of five million, they need partnership, policy, and funding. We sit with them in a meeting at the Ministry of Health. The government official is impressed but noncommittal. "This is excellent," he says, tapping a pile of papers on his desk. "But the budget has already been allocated for the next two years. We have to prioritise the new hospital wing. You should apply for a grant from an international foundation." The innovators leave the meeting visibly frustrated. Their problem is no longer technical; it is bureaucratic. Their brilliant, agile solution has collided with the slow, monolithic inertia of a system wedded to old models of centralised, infrastructure-heavy projects.

"*The world is not short of brilliant, life-saving innovations for climate adaptation. We are drowning in successful pilot projects. The bottleneck is not invention; it is implementation. It is the 'valley of death' between a proven local solution and a scaled-up, system-level program. This requires a revolution in how governments and large-scale funders identify, finance, and empower local innovators.*"

— Rajiv Shah, President, The Rockefeller Foundation.

The stories of Priya, Jean-Pierre, and the Afya Mama team prove that the solutions are within reach. Human ingenuity, when focused and directed, is more than a match for the challenges we face. We can power remote clinics, build cooler homes, and deliver lifesaving information. The technical adaptation gap is closing.

But their stories also reveal a deeper, more stubborn gap. The real adaptation gap is not a deficit of technology. It is a deficit of political will. It is a failure of our global systems of policy and finance, which are still captured by the inertia of the fossil fuel economy, to move at the speed and scale that the crisis demands.

The brilliance of a solar-powered blood bank in a remote Rwandan clinic is a beacon of hope. But its light also illuminates the larger darkness surrounding it: a global system that is failing to fund and scale these life-saving solutions. Ingenuity from the ground up is essential, but it is not enough. It must be met with a top-down policy revolution. The advocate's fight must now join the innovator's battle.

The workshop must now give way to the halls of power, where a global call to action is long overdue.

CHAPTER 16 | THE HUMAN BAROMETER

Some losses are not counted because they were never born. But they are felt — in the empty cribs, the buried ultrasounds, the unnamed grief that moves through mothers like a second heartbeat. These are the stories that rarely enter the data, the names that never reach a registry. Yet they are the truest record of a world in crisis. An archive written in blood and silence.

Barely five years ago, a microscopic entity brought our modern world to a standstill. The COVID-19 pandemic was a biological event, not a climatic one, yet for us — a doctor who has treated the sick and a scientist who models systemic risk — it is the most vital, terrifying, and instructive event of our lifetimes. It was a full-dress rehearsal for the era of climate shocks. It was a global fire drill that exposed the profound, shocking fragility of the very systems we depend on for our survival.

We all share the memory of that time. The eerie, unnatural silence of New York's Times Square, of the Roman Colosseum, of Mecca's Grand Mosque. The nightly news reports with their rising, relentless death tolls, presented in terrifying, data-rich graphics. The faces of exhausted, mask-bruised nurses, their eyes hollowed out with a grief and fatigue we could only guess at. The Crisis didn't just test our healthcare system; it also revealed the weakness in our social structures, economies and politics. The secondary impacts were a pandemic in themselves: the missed cancer screenings, the interrupted childhood immunisations, the catastrophic rise in domestic violence and mental health crises. The world's magnificent healthcare apparatus, built over a century, revealed itself to be a finely tuned instrument designed for a world of predictable problems, utterly unprepared for a true system shock.

For pregnant women, this shock was a unique and terrifying ordeal. We remember the story of a midwife named Esther, working in a sprawling peri-urban settlement near Cape Town. During the harshest lockdowns, her work became an exercise in clandestine heroism. We spoke to one of her patients, a young woman named Nomusa, who went into labour two weeks early. The local clinic had been converted into a COVID-only facility and was turning away all other patients. Taxis were not running due to the curfew. Nomusa's only lifeline was her cheap mobile phone. Huddled in her small shack, with her terrified husband as her only help, she relied on Esther's voice, a crackling, disembodied presence over a WhatsApp video call, to guide her through every contraction. Esther, from her own home miles away, had to diagnose a complication — a baby in the breech position — by watching the distorted image on her tiny screen. The successful birth was a miracle of individual courage and skill, but it was a miracle necessitated by the complete failure of the system around them.

Globally, the statistics confirm this anecdotal reality. A 2021 study reviewing data from 40 countries found that the pandemic was associated with a significant global increase in stillbirths, maternal deaths, and maternal depression. The reasons were a textbook illustration of system failure. Decades of slow, hard-won progress were reversed in a matter of months.

The COVID-19 pandemic was not a climate event. But as a doctor and an environmental scientist who have now studied both crises, we can tell you with absolute certainty that it is the most accurate and chilling blueprint we have for the future of maternal health in the climate era. The specific pathogen will be different, but the mechanisms of collapse it revealed are identical to those that will be triggered by major climate shocks. The pandemic provided us with a perfect, devastating preview of our vulnerabilities, which can be understood through four clear parallels.

The first is System Diversion. During the pandemic, the entire global health apparatus was forced to pivot to a single, overwhelming priority: fighting the virus. In India alone, during the first lockdown, it is estimated that over 25 million planned childbirths were disrupted.

Resources, staff, and political attention were all diverted, leaving "routine" services like prenatal care and safe delivery to collapse. In the climate era, the "virus" will be the disaster of the moment. When a mega-flood inundates a coastal region, the entire health system will be diverted to the crisis of trauma, displacement, and waterborne disease, just as it was in Malawi for Dr Mponda. The long-term burnout of healthcare workers from the pandemic has left this system with even less resilience to face the coming shocks.

The second parallel is Supply Chain Collapse. The pandemic revealed the shocking brittleness of our globalised, demand-driven supply chains, an economic model that prioritises efficiency over resilience. The scramble for masks and then vaccines, and the subsequent rise of "vaccine nationalism," exposed how quickly global cooperation breaks down under pressure. A major climate event will cause an even more profound collapse because it destroys the physical means of transport. It is a lesson in the "last mile" problem. An international aid flight might land with a planeload of midwifery kits in a nation's capital, but if the country's roads and bridges are washed out, those kits are useless. They will sit in a warehouse while women die just a hundred kilometres away.

The third parallel is Movement Restriction and Access Failure. The lockdowns of 2020 were a shocking lesson in how quickly our freedom of movement can be curtailed. Climate shocks will impose their own, more brutal lockdowns. A government can eventually lift a quarantine. No government can lift a bridge that has been washed away, a road that is buried under a landslide, or a landscape that is on fire. We saw this with the pregnant women in rural areas who had travelled to cities to await their delivery in "maternity waiting homes" and then found themselves trapped, far from their families and support systems, when lockdowns were imposed. This physical isolation from care will become a defining feature of the climate era.

The final, and perhaps most insidious, parallel is the Erosion of Trust. The pandemic unleashed a secondary, equally dangerous epidemic of misinformation, fueling a deep and damaging mistrust of science, public health institutions, and governments. This erosion of social cohesion makes a coordinated response to any crisis infinitely more difficult. The repeated, chaotic, and often unequally distributed impacts of climate disasters will do the same, deepening political polarisation and making it harder for communities to come together. The political fracturing we saw during COVID is a gentle preview of the conflicts that will arise when the resource at stake is not a vaccine, but drinking water and food.

"*COVID-19 was a stress test for our global health security systems, and we failed. The weaknesses it exposed — in supply chains, in equitable access, in international cooperation, and in public trust — are precisely the vulnerabilities that will be exploited by the accelerating crisis of climate change. We have been warned.*"

— Dr Michael Ryan, Executive Director, WHO Health Emergencies Programme.

Some will argue this comparison is flawed. A pandemic is a singular biological event, whereas climate change is a chronic, multi-faceted environmental process. This is true, but it misses the crucial point. The future of the climate crisis will not be one slow, steady decline. It will be a future of repeated, stuttering, and overlapping acute shocks. The real challenge is that the enemy is invisible and ubiquitous. A virus can be isolated, sequenced, and targeted with a vaccine. How do you fight an enemy that is also the very air you breathe, the rain that falls from the sky, and the economic system that powers your world?

Furthermore, the climate crisis is, in one crucial respect, more dangerous. For COVID-19, we eventually developed treatments and vaccines. There is no vaccine for a flood. There is no antiviral for a drought. There is no monoclonal antibody that can stop a wildfire. The "disease" of the climate crisis is not an external pathogen we can target; it is a chronic condition built into the very DNA of our global energy and economic systems. The treatment is not a shot in the arm; it is a complete transformation of the way we live.

The ghost of the pandemic is therefore not a memory, but a prophecy. It was our fire drill. It was a non-fatal, global system shock that gave us a perfect preview of our vulnerabilities. We saw, with our own eyes, how quickly progress can be undone. We all talked about "building back better." But the truth is, we have largely failed to learn the lesson. We have primarily used the immense resources of the recovery to restore the very same fragile, centralised, just-in-time systems that broke so spectacularly. The desire for a return to "normalcy" has led to a collective amnesia about how abnormal things became, and how quickly. We didn't just fail the test; we are now actively trying to forget the test ever happened.

This failure forces us to confront a fundamental question. If our current systems are all but guaranteed to fail in the face of the coming shocks, then what is our measure of progress? What is our definition of resilience? The metrics we have used for a century — GDP growth, stock market performance — are not just inadequate; they are actively misleading, encouraging us to rebuild in the same vulnerable ways.

If we are to survive the coming shocks, we need a new way to measure what truly matters. We need a new barometer for resilience, one that is not based on money or markets, but on the undeniable, fundamental truth of human life. And if we are to understand the depth of what is vanishing, we must listen not just to the mothers — but to the girls becoming mothers in a world they didn't break, but must now repair.

CHAPTER 17 | THE HUMAN BAROMETER

Before they are mothers, they are daughters. Before they are protected, they are sent for water, for firewood, for anything the village needs and the state has forgotten. In many places, girlhood is not a season — it is a trial. And in a world growing hotter and harsher, these girls do not inherit innocence. They inherit the fire.

Night has fallen at the Zambezi's edge. The visual spectacle of the Falls, with its rainbows and roaring mist, has been swallowed by the immense darkness of the African continent. But the sound remains. It is the first and last thing you notice here, a constant, rhythmic, and mighty roar that vibrates through the ground, up through the soles of your feet, and into your very bones. It is a planetary heartbeat, a ceaseless reminder of the raw, elemental forces that govern this world.

We began this journey together with a question: In an age of escalating crisis, how do we

Measure what truly matters? For half a century, we have tried to chart our course through the climate crisis using a set of familiar but deeply inadequate metrics. We measure our failure in parts per million of carbon dioxide, a language of atmospheric chemistry. We measure the stakes in degrees Celsius, a language of abstract thermal dynamics. We measure the damage in billions of dollars, the cold, amoral language of insurance ledgers and economic models. We have celebrated a 3% rise in global GDP in the same year that saw the most significant single-year increase in wildfire smoke exposure in human history, condemning an unknown number of pregnancies to the risks we saw in California — one metric registered success; the other, a silent biological catastrophe. Our instruments are broken.

These metrics are scientifically valid, but they are emotionally and morally sterile. They are the language of spreadsheets, not of life. They tell us the planet has a fever, but they fail to capture the agony of the patient. After the journey we have taken, we are here to propose a new metric. We need a new way to measure our progress or our failure, a new vital sign that is immediate, undeniable, and universally understood.

We propose the Human Barometer: the health and survival of the most vulnerable mother and her newborn child at the moment of birth. We propose that this single, intimate event is the most powerful and accurate indicator we have for the overall health of our societies and our planet. It is the ultimate metric. If she and her baby are safe, healthy, and supported, it means the systems that sustain us all are functioning. If they are not, it means we are failing in the most fundamental way possible.

Consider the evidence we have gathered, the stories that are now woven together. We met Eliza, giving birth in the mud as Cyclone Freddy tore her world apart. Her story is a measure of a system's resilience — or fragility — in the face of an acute shock. Her ordeal was the price of our collective failure of imagination, our inability to conceive of a storm like Freddy.

We walked with Zara on her long, desperate journey and stood with the midwife Neïma Patricia as she fought to create a sanctuary of life. Their world is a measure of a system's response to a chronic crisis. The health of the mothers in Neïma's care is a direct, daily referendum on the slow, grinding impacts of drought and conflict. We can celebrate Neïma's stunning success,

But we must also confront the profound absurdity that in an age of artificial intelligence, the survival of a mother and child in the Sahel still depends on the heroic efforts of one woman in a tent — a woman who has now heard stories of similar struggles from midwives in flooded Malawi and knows her fight is not unique.

And we sat with Claire in a high-tech Californian hospital, her sense of safety shattered by the revelation that the smoke from a past wildfire had become a ghost in her own body, her doctor connecting her condition to the same patterns of heat stress seen in women like Jessica in Phoenix. Her story is a measure of our perceived immunity. It proves that the firewall is a myth.

Eliza, Neïma, and Claire live in vastly different worlds. But the planetary fever connects them. Their stories are not parallel; they are convergent. They ask the same question: in the face of this crisis, can we protect the sacred, vulnerable process of birth itself?

The Human Barometer is the ultimate metric for four fundamental reasons.

First, as Dr Robinah knows from a lifetime of practice, it is holistic. A safe and healthy birth is the successful outcome of an entire system of systems. For a mother and baby to thrive, she must have had nutritious food and clean water (a functioning ecosystem), a place of safety (a functioning society), access to care (a functioning health system), and the support of her community (a functioning culture). If the mother and baby are well, it tells you that everything — from the soil to the soul of the community — is also, in some essential way, well. It is the ultimate measure of intergenerational equity: a society that cannot guarantee the safe passage of its next generation into the world has, in effect, defaulted on its debt to the future.

Second, as Zebedee knows from his work modelling complex systems, it is predictive. Just as a cardiologist looks at inflammation markers in the blood to predict a future heart attack, the Human Barometer allows us to see the 'social inflammation' that precedes state failure. A rise in stillbirths or low-birth-weight babies in a region is a definitive sign that the entire system is under profound stress, often before other, slower indicators like GDP contraction or mass migration are officially recorded. It is the canary in the coal mine.

Third, it is moral and unifying. Unlike a metric like GDP, which is abstract and morally neutral, no one of decent conscience can argue against the value of a mother's and a child's life. This metric cuts through the sterile, polarising debates over economic trade-offs. It is a powerful tool against the obfuscation of the vested interests of the fossil fuel economy, who seek to muddy the waters with economic complexities. It is much harder for them to argue against the moral imperative to save a baby's life. It reframes the entire climate debate from one of cost-benefit analysis to one of fundamental, shared human values. It is a metric rooted not in economics, but in love.

Finally, it is actionable. The Human Barometer is not just a philosophical concept; it drives a practical, clear agenda. If your primary goal is to improve the health and survival of mothers and newborns, it forces you to invest in the very things that build true, foundational resilience: clean water, local food systems, women's education, and decentralised, solar-powered community health clinics. It provides a clear, practical roadmap for investment that builds a better, more equitable world from the ground up. So, what now? What does this new barometer demand of us?

It demands that we change the questions we ask. The call to action is not just for the leaders at global summits, but for all of us, in our own spheres of influence. For the policymaker, it is to ask of every new law not just, "How will this affect GDP?" but "How will this affect the health and safety of the most vulnerable mother and child?" For the engineer, the question is, "How can my design function when the grid fails?" For the doctor or nurse, it is to ask, "How can I become an advocate for the environmental conditions that are shaping my patients' health?" For the citizen, it is to ask, "Am I supporting a system that values profit over placentas?"

The journey we have taken through this book has been a harrowing one. It was a diagnosis of a planetary disease, a witnessing of its most painful symptoms. But the purpose of a diagnosis is not to pronounce a death sentence. It is to create a treatment plan. The fight for the future is not a fight to restore the old, fragile world, which was already failing so many. It is a fight to build a new one, a world that is genuinely resilient because it is more just, more equitable, and more aligned with the fundamental systems of life.

The roar of the water here at the Zambezi's edge is constant, a reminder of a power far greater than our own. It is the sound of the world as it is. But there is another sound, the sound we have been fighting for throughout these pages. It is the first cry of a newborn child — a sound of defiant, vulnerable, and sacred life. It is the sound of the world as it could be. The ultimate purpose of all our work, all our policies, all our technology, and all our love is to ensure that this sound is one of joy and hope, for every mother, everywhere.

That is the only test that matters. That is the promise of the Human Barometer.

THE MANIFESTO: THE HUMAN BAROMETER PLEDGE

The Human Barometer — A Toolkit for the Future

This is not the end of the story. It is the moment we place the stethoscope back on the chest of the world and listen again — not just to the data, but to the pulse beneath it…

The diagnosis has been made. The chart is full. The signs are clear. The question now is no longer "What's wrong?" but "What will we do with what we know?"

We have walked through fire and famine, flood and flight. We have listened to the silence between contractions and the roar of systems collapsing. We have stood beside women whose lives mark the frontline of a planetary pathology. This is not a metaphor. It is medicine. And every medicine needs a response plan.

So here, in the spirit of the midwives and mothers who held the line when systems failed, we offer not solutions, but instruments. This is not a comprehensive map. It is a toolkit — a set of vital signs, hard-won practices, and new reflexes for an age of cascading emergencies.

1. Listen Beyond the Noise

The earliest signs of collapse are often inaudible to systems built to track only what screams. Learn to hear what the fevered body hears: the flutter before the silence, the tremor beneath the policy.

> Vital question: Who is already whispering what the data will confirm too late?

2. Disaggregate the Danger

Global averages conceal local disasters. A two-degree rise means nothing to a spreadsheet and everything to a mother in a tin-roofed home at 42°C. Climate vulnerability is not equal; it is deeply gendered, racialised, and political.

> Vital question: Who bears the heat, the distance, the silence — first and most?

3. Name the Losses That Don't Make the News

Stillbirths in heatwaves. Anaemias that bloom in famine. Girls were pulled from school to fetch water farther away. These are not "downstream" effects. They are central indicators of collapse. Count what matters, not just what markets can price.

> Vital question: What lives have we rendered invisible by refusing to measure them?

4. Build Trust Like Infrastructure

Hospitals collapse in days. Trust takes years. But in crisis zones, trust can save more lives than supply chains. Invest in those who are already trusted — community midwives, women's networks, faith leaders who walk the talk.

> Vital question: Who already has the ears — and the consent — of the people we hope to serve?

5. Shift the Epicentre of Expertise

Expertise does not always sit in Geneva or New York. It walks barefoot in Kalambari. It cooks on firewood in Jackson. It catches babies in tents and diagnoses danger by scent, by pulse, by pattern.

> Vital question: What knowledge have we mistaken for marginal, simply because it was born in survival?

6. Treat Climate as a Clinical Emergency

Not a distant scenario. Not an economic risk. A crisis of bodies, systems, and care. Every heatwave is a mass casualty event waiting to be measured. Every flooded clinic is a triage site. Every maternal death in a heat dome is a verdict on our readiness.

> Vital question: If this were your hospital, would you still call it adaptation?

7. Act Like the Future Is Already Here — Because It Is

Kalambari is not an exception. It is a preview. Jackson is not an outlier. It is a mirror. We must stop mistaking these places for the edge. They are the epicentre of what is coming — and what is already here.

> Vital question: What would you do differently if you believed these women were not far away, but next?

The Final Sign

This book is not a lament. It is a ledger. A record of symptoms misread, and lives miscategorised. But it is also a prescription.

To policymakers: Do not wait for metrics to catch up with mortality.

To clinicians: See the climate in your patients' bloodwork, their wounds, their breath.

To scientists: Let your graphs speak names.

To all of us: Listen like the roof is already lifting.

The human barometer is not an idea. It is not a symbol. It is a body.

A body under siege. A body still birthing. A body demanding a new kind of attention. We are not at the beginning of the crisis.

We are in the middle of the diagnosis.

The only question that remains is whether we are brave enough to treat the patient before the chart goes still.

CODA — Listening Again

The last sound a healer learns is silence.

Not the sterile hush of a ward at midnight, but the deep, breathing quiet that comes when a patient no longer fights the rhythm of the world.

In that stillness, a pulse becomes more than a signal. It becomes a language.

We began this journey listening for the murmurs inside a single chest, tracing the subtle tremor that warned of fever, the whisper in the blood that spoke of collapse. But what we found was a resonance larger than one body. The same sound that threads through a mother's heartbeat travels through the soil, through the ocean's lungs, through the atmosphere that keeps us all alive.

The Human Barometer was never meant to measure crisis. It was meant to teach us to hear again. To recognise that care is not a policy, and progress is not a cure. That every data point, every forecast, every desperate cry in the dark is part of a single, trembling note: the plea of a world that still wants to live.

One day, the instruments will fail — the sensors will dim, the satellites will fall quiet — and what will remain is the one tool no machine could replace: the human ear pressed gently against the world, listening for a heartbeat. If we can hear it, truly listen to it, we may still learn how to keep it beating.

About The Author

The Human Barometer is the groundbreaking result of two distinct worlds of expertise colliding to reveal a single, terrifying truth.

Dr Robinah Namulindwa is the book's clinical heart. A Ugandan-born gynaecologist and maternal health advocate, she is the founder of Clinics for Maternal Healthcare in Rural Africa. Dr Namulindwa brings decades of frontline experience from the "examination room," witnessing firsthand how the planet's "fever" and "withered fields" manifest as a life-or-death crisis in the bodies of pregnant women and their newborns.

Zebedee Njisuh Feka is the book's planetary diagnostician. A distinguished environmental scientist and coastal resilience specialist, he has conducted extensive research on ecological collapse, resource scarcity, and climate shocks across West-Central Africa, providing the searing "why." He traces the vital signs of the entire planetary system, from the "incredible shrinking lake" to the "endless cyclone".

Together, this doctor-and-scientist team connects the fever chart of a pregnant woman to the temperature graph of the Earth. They expose the "glaring omission" in global policy, proving with devastating clarity that the health of a mother and child is the only true measure of our planet's survival.

www.ingramcontent.com/pod-product-compliance
Lightning Source LLC
LaVergne TN
LVHW022224200926
848481LV00006B/2104